310

95

LIST OF CHANGES IN BRITISH WAR MATERIAL

IN RELATION TO

EDGED WEAPONS, FIREARMS

AND

ASSOCIATED AMMUNITION AND ACCOUTREMENTS

With Alphabetic Index

VOLUME III
1900–1910

Compiled by
Ian D. SKENNERTON

This edition copyright 1987
by I. D. Skennerton

and

Published 1987 by
Ian D. Skennerton
Box 56 P.O.
Margate 4019
Australia

ISBN 0 9597438 5 5

How to use the List of Changes

The List of Changes were issued monthly with Army Circulars in numerical sequence, commencing with No. 1 in 1860. Published by the War Department, the L.o.C.'s advised of changes in all military stores, ranging from bayonets to bicycles, hospital furniture to howitzers, and ammunition to admiral's chamber pots.

When first introduced they were known as "Changes in Artillery Matériel, Small Arms, Accoutrements and other Military Stores", and in the late 1860's, became "List of Changes in Artillery Matériel, Small Arms and other Military Stores". In 1872, they were renamed "List of Changes in War Matériel and of Patterns of Military Stores". For the sake of convenience, they are often referred to just as "List of Changes", "L.o.C.'s", or even just "L.C.'s".

The official date of introduction of an article into service is usually that of the Army Circular—a table with these numbers and dates will be found on the following page. The approval date of an item usually differs from that of its introduction, sometimes even by a number of years, and yet other items were never included in the List of Changes; such as the Martini-Metford rifles, and the Martini-Enfield Cavalry Carbine Mark II. Some items, though announced in these Lists, were occasionally never even issued. Articles may be found to have more than one approval date: this is because there was more than one subject involved, and the different approval dates refer to seperate items, or parts thereof.

From late 1891, the letters "L", "N" or "C" often appear near the title or heading—such letters refer to the service involved, these being "L" for Land,
"N" for Naval
and "C" indicating Common to both services. The original Lists use the symbol "§" to indicate "List of Changes, number", though the abbreviation "L.o.C." is used in this volume.

Approval dates of the individual items will be found on the right hand side of the item designation, and the designations are often followed by a further description of the article, e.g.

10258–Frog, buff, sword.   ·   **L**   **5 Jun 1900**
7¾-inch, with loop; Pioneers and R.A.M.C.

Abbreviations may be encountered, and include:

A.O.D.	Army Ordnance Dep't	O.R.	Other ranks
C.O.O.	Chief Ordnance Officer	R.A.	Royal Artillery
G.S.	General Service	R.A.M.C.	Royal Army Medical
H. & F.	Horse & Field	R.F.	Rimfire Corps
M.E.	Martini-Enfield	R. & F.	Rank & File
M.L.E.	Magazine Lee-Enfield	S.	Serjeant
M.L.E.R.	Magazine Lee-Enfield Rifle	S.S.	Staff-serjeant
M.L.M.	Magazine Lee-Metford	W.O.	Warrant Officer
M.M.	Martini-Metford		

Also refer to Vols. I & II.

Year	Month	Range	Year	Month	Range	Year	Month	Range
1900	Feb	9981-10008	1904	Jan	11946-11991	1908	Jan	13968-13997
	Mar	10009-10039		Feb	11992-12056		Feb	13998-14045
	Apr	10040-10120		Mar	12057-12089		Mar	14046-14071
	May	10121-10158		Apr	12090-12128		Apr	14072-14107
	Jun	10159-10188		May	12129-12182		May	14108-14142
	Jul	10189-10219		Jun	12183-12219		Jun	14143-14178
	Aug	10220-10256		Jul	12220-12267		Jul	14179-14207
	Sep	10257-10284		Aug	12268-12325		Aug	14208-14237
	Oct	10285-10312		Sep	12326-12388		Sep	14238-14287
	Nov	10313-10341		Oct	12389-12420		Oct	14288-14330
	Dec	10342-10369		Nov	12421-12468		Nov	14331-14367
				Dec	12469-12524		Dec	14368-14411
1901	Jan	10370-10390	1905	Jan	12525-12566	1909	Jan	14412-14431
	Feb	10391-10417		Feb	12567-12608		Feb	14432-14477
	Mar	10418-10437		Mar	12609-12649		Mar	14478-14528
	Apr	10438-10470		Apr	12650-12705		Apr	14529-14574
	May	10471-10522		May	12706-12761		May	14575-14612
	Jun	10523-10556		Jun	12762-12823		Jun	14613-14640
	Jul	10557-10594		Jul	12824-12857		Jul	14641-14686
	Aug	10595-10636		Aug	12858-12892		Aug	14687-14717
	Sep	10637-10685		Sep	12893-12950		Sep	14718-14765
	Oct	10686-10720		Oct	12951-12989		Oct	14766-14794
	Nov	10721-10761		Nov	12990-13019		Nov	14795-14841
	Dec	10762-10796		Dec	13020-13055		Dec	14842-14880
1902	Jan	10797-10858	1906	Jan	13056-13104	1910	Jan	14881-14915
	Feb	10859-10916		Feb	13105-13134		Feb	14916-14944
	Mar	10917-10973		Mar	13135-13189		Mar	14945-14978
	Apr	10974-11031		Apr	13190-13230		Apr	14979-15008
	May	11032-11077		May	13231-13271			
	Jun	11078-11107		Jun	13272-13322			
	Jul	11108-11148		Jul	13323-13360			
	Aug	11149-11201		Aug	13361-13397			
	Sep	11202-11264		Sep	13398-13442			
	Oct	11265-11287		Oct	13443-13474			
	Nov	11288-11328		Nov	13475-13520			
	Dec	11329-11366		Dec	13521-13556			
1903	Jan	11367-11413	1907	Jan	13557-13580			
	Feb	11414-11459		Feb	13581-13618			
	Mar	11460-11496		Mar	13619-13653			
	Apr	11497-11536		Apr	13654-13684			
	May	11537-11601		May	13685-13712			
	Jun	11602-11657		Jun	13713-13751			
	Jul	11658-11712		Jul	13752-13786			
	Aug	11713-11767		Aug	13787-13828			
	Sep	11768-11806		Sep	13829-13862			
	Oct	11807-11847		Oct	13863-13890			
	Nov	11848-11894		Nov	13891-13939			
	Dec	11895-11945		Dec	13940-13967			

10030—Box, ammunition, small-arm, Mark XIII.   L   23 Dec 1899

No more to be made.

Nomenclature.

No more of the above-mentioned boxes (LoC 6153) will be made, and when the existing stock is used up, the pattern will become obsolete. The nomenclature has been altered as shown above.

10042—Sword, practice, gymnasia,   L   20 Feb 1900
converted 1895 pattern.

A pattern of the above-mentioned sword has been sealed to govern the conversion of swords, practice, gymnasia, pattern 1895 (LoC 7848) as may be ordered.

The conversion consists in reducing the blade in width and thickness to assimilate it as nearly as possible to the pattern 1899 practice sword (LoC 9984).

Weight of converted sword . . . . 1 lb. 11¼ oz.
Balance from hilt. . . . . 1½ inches.

10094—Cartridges for instruction: Small-arm.   C   19 Oct 1899

In future manufacture, cartridges of the above-mentioned description, which are issued for instructional purposes (LoC 8955), will be pierced with two holes at right angles through the case, as shown in the annexed typical drawings, so that it can at once be seen that the cartridge is not a service one.

Existing cartridges will be dealt with as follows:—
Cartridges with solid-drawn cases- To be altered locally, and on board Her Majesty's ships, in accordance with the instructions given below.
Cartridges with rolled-brass or paper cases- To be returned to the Royal Arsenal, Woolwich, for alteration.

Instructions for carrying out alteration.

Mark the position of the centre of the hole, and drill a small hole by means of a brace of any convenient size, open out the hole with a round file and finish it with a rimer of the required size. The diameters of the holes are to be: Machine gun below 1-inch calibre and small-arm (except Webley pistol)— 3/16-inch. The holes near the base are to be just above and clear of the cap chamber, the upper holes being midway between the bottom holes and the neck of the case.

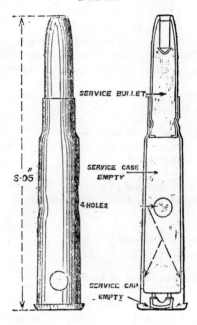

LoC 10094

10098–Boxes, ammunition, S.A.   C   18 Jun 1895
                                              16 Apr 1895
Linings to be labelled   20 Mar 1900

Boxes of the above-mentioned description will, in future, have a label (similar to that placed on the outside of the box) affixed to the centre of the closing plate. The particulars of the batch of cordite, and the number of the box, will be stamped on the closing plate after it is soldered down.

10117–Cartridge, signal, Very's, white. (Mark III) N   23 Feb 1898

A design of "cartridge, signal, Very's, white" has been adopted for naval service, and a specification has been approved to govern future manufacture.

The cartridge (which will be known as Mark III) is generally similar to the green and red cartridges described in LoC 8930. The rim of the head of the cartridge is milled half way round, the remainder being left smooth. The paper portion of the cartridge projecting beyond the brass case, and the cardboard disc, are white. Dimensions are the same as in LoC 5173.

10122—Bottle, oil, Mark III   C   19 Dec 1899
    Brass, all .303-inch magazine rifles and carbines, with washer and stopper.

Stopper   C
With spoon, 1 per bottle.

Washer   C
Leather, 1 per bottle.

A pattern of the above-mentioned oil bottle, of which a drawing is attached, has been sealed to govern future manufacture and the alteration of existing oil bottles of previous pattern.

It differs from the Mark II oil bottle (LoC 6761) in having a larger mouth, to make it suitable for mineral jelly (LoC 10131), and a flat-headed stopper with spoon to facilitate the removal of the mineral jelly from the bottle when required for use.

The altered bottle is suitable for rifle oil, and will be used therewith until the stock of rifle oil is used up.

Full size.

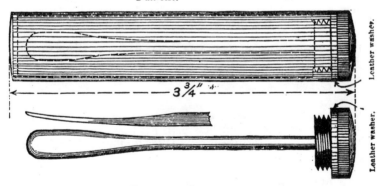

Stopper with spoon.

10123—Mandril, scabbard, sword, cavalry, pattern 1885, Mark II.   19 Mar 1900

Extended use.

With reference to LoC 6865; the above-mentioned mandril can be used if necessary with scabbards, sword, cavalry, pattern 1899, new and converted, described in LoC 9880.

10131—Mineral jelly. C 11 Nov 1899
16 Mar 1900

To be used in place of rifle oil, for .303-inch magazine rifles, carbines and machine guns.

In future, mineral jelly will be used, in substitution for rifle oil (LoC 7468) for cleaning and oiling the barrels and exterior parts of all .303-inch magazine rifles and carbines and the barrels of .303-inch machine guns, and for all purposes for which rifle oil has hitherto been used.

The existing store of rifle oil will be used up.

Rangoon oil will still be used for oiling the breech mechanism of magazine rifles and carbines and the working parts of .303-inch machine guns as formerly.

10160—Gauges armourers', pistol, Webley— 16 May 1900
 Gauge— distance of cylinder from face of body, rejecting, Marks I, I*, II and II* pistols.

Obsolete.

With reference to LoC 9832— As the .049-inch gauge described therein has been found to be suitable for all marks of Webley pistols in the service, the .047-inch gauge will be regarded as obsolete.

10190—Pistols, Webley, Marks I, I*, II and III. C 19 Oct 1899

Repair.
LoC 9832 amended.

No more Mark II hammers or Mark I triggers will be made, and as soon as existing stock of these parts has been used up, Mark IV hammers and Mark IV triggers will be used for the repair of all marks of Webley pistols.

At present the ratchets of Mark IV cylinders only are hardened. Whenever Mark III pistols require new cylinders, Mark IV cylinders will be fitted thereto, and at the same time, pawls with hardened points should be fitted. Where a Mark I cylinder is used in repair, (i.e. in the case of Marks I, I* and II pistols) a pawl with unhardened point, if obtainable, should be used.

The Mark (numeral) of pistols of old pattern fitted with new parts will not be altered. All references to Marks II* and III* pistols in LoC 9832 will be removed.

10191—Scabbard, sword-bayonet, pattern 1888.   L   24 May 1900
    Brown leather.

A pattern of the above-mentioned scabbard has been sealed to govern manufacture as may be ordered.

The scabbard differs only from the "Scabbard, sword-bayonet, pattern 1888, Mark I" described in LoC 5877, in the leather being left brown instead of being stained black.

10220—Carbine, magazine, Lee-Enfield.   L   (no date)
    Fitted to take the pattern 1888 sword-bayonet.

A pattern of the above-mentioned carbine has been sealed to govern future manufacture.

It differs from the Carbine, Lee-Enfield, Mark I (LoC 8390), and the Mark I* (LoC 9700), in the following manner—

The barrel is special, and is increased in diameter at the muzzle to fit the sword-bayonet. The sighting is the same as the Martini-Enfield artillery carbine, Mark III (LoC 9786). The fore-end is special and fitted with nose-cap, Martini-Enfield artillery carbine. The handguard is also special, and fitted with two springs.

    Length of barrel . . . . .21 ins.
    Calibre . . . .303-inch.
    Rifling. . . . . .Enfield system.
    Grooves, number. . . . . .5.
    Grooves, depth . . . .005 inch.
    Width of lands . . . .0936 inch.
    Spiral (left-handed) . . . . .1 turn in 10 ins.
                                                   or 33 calibres.
    Sighting (system). . . . . .V-notch and bar-
                                                   leycorn foresight.
    The foresight is fitted with wings for its protection.

    Length of carbine without sword-bayonet . .3 ft. 4¼ ins.
    Length of carbine with sword-bayonet. . . . .4 ft. 4¼ ins.
    Weight of carbine . . . . .7 lb. 8 oz.
    Weight of carbine with sword-bayonet. . . . .8 lb. 7½ oz.

10221—Hook, suspending sword-bayonets,   L
    (Mark I).
    Brass; for suspending sword-bayonets from
    rifles, and carbines which take sword bayonets,
    in new pattern arms racks.

A pattern of the above has been sealed to govern future manufacture and supply.

The hook is intended to be used for attaching the sword-bayonets to the rifles, and carbines which take sword bayonets, when standing in the arm racks in store.

The sword-bayonet with scabbard is passed through the wide end of the hook, and suspended to the rifle by passing the narrow end through the band swivel.

10258—Frog, buff, sword.   L   5 Jun 1900
    7¾-inch, with loop; Pioneers and R.A.M.C.

Alteration.

To enable this frog (LoC 3994, Fig. 8) to carry the Martini-Henry converted sword-bayonet, the slit in the front will be lengthened 3/16-inch.

Troops in possession of this pattern sword-bayonet will have their frogs altered regimentally.

10273—Cartridge, S.A., ball, pistol, Webley,   C   14 Jul 1900
    cordite, (Marks II and III).   8 May 1900
    Also Enfield.   27 Jun 1900

No more Mark III to be made.
Manufacture of Mark II to be reverted to.
Addition of glazed board disc.
Packing of Mark II cartridges.

With reference to LoC 9159— No more of the above-mentioned Mark III cartridges will be made, and the manufacture of Mark II will be reverted to.

Existing stock of Mark III cartridges in the land service will be used up for practice purposes at home, the stock of Mark III at stations abroad being sent home on receipt of supplies of Mark II from Woolwich. All service and mobilization pistol ammunition at home should be exchanged for Mark II.

Mark III ammunition in the naval service will be withdrawn in accordance with arrangements which will be made by the Storekeeper-General of Naval Ordnance.

In future manufacture of Mark II, a glazed board disc (.02-inch thick) will be placed under the bullet on the top of the cordite. This alteration will not involve any change of numeral. A few Mark III cartridges have been issued with this modification.

Packing of Mark II cartridges.

Land Service
Box, ammunition, S.A. pistol, Enfield, 240 rounds.
276 rounds in box; 12 rounds in bundle.
Weight of filled box— 16 lb. 8 oz.

Naval Service.
Box, ammunition, S.A., half, naval.
828 rounds in box; 6 rounds in bundle.
Weight of filled box— 47 lb. 8 oz.

10342—Mineral Jelly.   19 Sep 1900
Bottles, zinc, oil.

With reference to LoC 10131— Troops equipped with the zinc oil bottle will continue the use of rifle oil instead of mineral jelly, this bottle not being adapted for containing the latter substance.

When the existing stock of zinc oil bottles has been used up, the Mark III oil bottle (LoC 10122) will be issued in its place, and mineral jelly will then be used.

10371—Rifle, M.L.E., Marks I & I*   C   22 Oct 21900

Alteration of sights.

With reference to LoC 8117 & 9700— The sights of the abovementioned rifles will, as a temporary measure, be altered as shown below.

The alterations and identifying marks will be as follows—

Method (1) for future manufacture.

Barleycorn of front sight solid, .02 inch to the left of axis of bore.

Method (2) for rifles sent to Birmingham for repair.

Barleycorn of front sight removeable, .02 inch to the left of axis of bore, and pinned into the block as in the M.L.M., Mark I* rifle (LoC 6760).

Rifles dealt with in accordance with methods (1) and (2) will be identified by a star (*) marked on the left hand side of the block of the foresight.

Method (3) for rifles in store or in the hands of troops.

The V notches in slide and cap of backsight .03 inch to the left of centre.

Rifles dealt with in accordance with method (3) will be identified by a star (*) marked on the face of the leaf at the right hand bottom corner.

10392—Action, skeleton, pistol, Webley. C 24 Jul 1900
3 Dec 1900

A pattern of the above-mentioned article has been sealed to govern conversion of such Webley pistols (various marks) as are not worth further repair as Service weapons.

10393—Rifles, M.L.E., Marks I and I*. C 22 Oct 1900

Alteration of sights.

LoC 10371 is hereby cancelled, and the following substituted:

With reference to LoC 8117 & 9700— The following alterations have been authorized to the sights of the above-mentioned rifles. The alterations and identifying marks to be as follows—

1. For future manufacture— Barleycorn of front sight solid, .02 inch to the left of axis of bore.

2. For rifles sent to Birmingham for repair— Barleycorn of front sight removeable, .02 inch to the left of axis of bore, and front sight removeable, .02 inch to the left of axis of bore, and pinned into the block as in the M.L.M., Mark I* rifle (LoC 6760).

Rifles dealt with in accordance with 1 and 2 will be identified by a star (*) marked on the right side of the block of the foresight.

3. Certain rifles in store and in the hands of troops have been altered as follows— The V notches in slide and cap of backsight .03 inch to the left of centre.

Rifles dealt with in accordance with 3 will be identified by a star (*) marked on the face of the leaf at the right hand bottom corner.

10394—Tools, armourers—
 Mandril, bayonet socket, Martini-Enfield L 18 Dec 1900

A pattern of the above-mentioned mandril has been sealed to govern manufacture.

It is to be used for the repair of Martini-Enfield bayonets when sockets become indented.

10419—Sword, naval, pattern 1900. N 20 Mar 1900
   Steel hilt, 28 inches.   17 Jan 1901

   Scabbard, sword, naval, pattern 1900 N 18 Apr 1900
   Black leather, steel mounted.   28 Dec 1900
   Also suitable for sword, naval, pattern 1889.

Patterns of the above-mentioned articles have been sealed to govern future manufacture.

The sword differs from the pattern 1889 (LoC 5848) in the blade being stiffer and fullered, and the grips being of leather, chequered, instead of iron.

   Length of sword from point to shoulder . . .28 inches.
   Length of sword over all . . . . .33.3125 inches.
   Weight of sword . . . . .2 lb. 9½ oz.
   Balance from hilt. . . . . .4.125 ins.

The scabbard differs from the pattern 1889 (LoC 5848) only in the mouthpiece being larger to take the stiffer blade.

   Length of scabbard . . . . .29 inches.
   Weight of scabbard . . . . .11 oz.

The pattern 1889 scabbard will become obsolete as the stock is used up.

10439—Rifle, magazine, Lee-Metford. C 25 Feb 1901
   (Marks II and II*)

   Alteration of backsight when fitted with Lee-Enfield barrels.

With reference to LoC 10393— Rifles, magazine, Lee-Metford, Marks II and II*, when requiring new barrels, may be repaired with barrels and bodies, magazine, Lee-Enfield, Mark I. When so repaired, they will require to be provided with a backsight having notches .03 inch left, unless there is a (*) marked on the right hand side of the block of the foresight, in which case a backsight with central notches is used.

Backsight with notches .03 inch left may be recognized by a star (*) marked on the face of the leaf at the right hand bottom corner.

10440—Belts, shoulder, brown, sword, L 28 Dec 1900
   "Sam Browne", pair. (Mark I).
   Leather, with two billets.

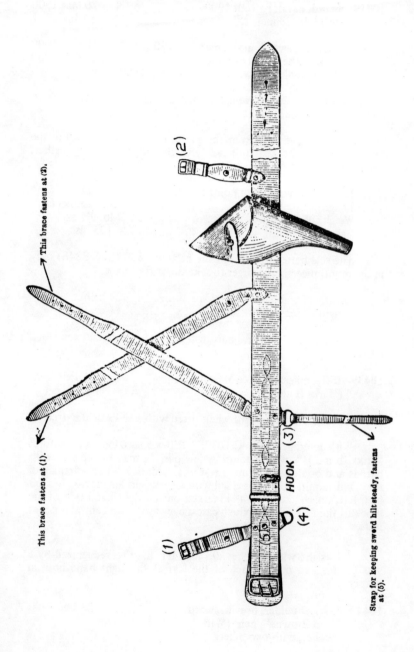

Belt, waist, brown, sword, "Sam   L
Browne", (Mark I).
Leather, with buckle, frog, and
21-inch by 5/8-inch straps.

Case, brown, pistol, "Sam Browne",   L
(Mark I).
Leather, with cover.

Pouch, ammunition, brown pistol,   L
"Sam Browne". (Mark I).
Leather.

Patterns of the above-mentioned articles, as shown in the accompanying drawings, have been sealed to govern manufacture.

The shoulder belts consist of two plain straps crossed at the back. Each strap is fitted at one end with a tongue, which passes through a dee on the back of the waistbelt, and is then secured to itself by a stud; the other end is buckled to a chape, which is secured to the front of the waistbelt as described above for the tongue.

The waistbelt is fitted with a double-tongued brass buckle and four brass dees, to which the ends of the shoulder belts are secured. There are also two brass rings for the attachment of the frog. The frog is fitted with two straps and studs, by which it is secured to the rings on the lower part of the belt, and with a 5/8-inch strap, which is attached to the rear ring of the frog to steady the hilt of the sword.

The pistol case is of the usual shape, and is fitted with a loop on the back to admit of its being carried on the waistbelt. A small hook is riveted to the loop, and passes through the belt to keep the case in its proper position.

The ammunition pouch is fitted with a collapsible gusset and a loop on the back to carry it on the waistbelt.

10441—Lockets, union, brass—   12 May 1900
    Royal Marines. (Mark II).   L   5 Nov 1900
    S. and R. and F.; lion, crown and
    regimental motto.

    Universal. (Mark II).   L
    S. and R. and F.; lion, crown, and royal motto.

    Lockets, union, gilt—
    Royal Marines. (Mark II)   L.
    Lion, crown and regimental motto.

Universal. (Mark II)   L
W.O. and S.S.; lion, crown and royal motto.

Patterns of the above-mentioned lockets have been sealed to govern future manufacture.

They differ from the previous patterns (not published in List of Changes) in the design of the Royal crest, which was not quite correct in the Mark I lockets.

10442–Sling, rifle, web, G.S. (Mark I).   L   31 Jan 1901
      Also R.A. carbine, 46 inches by 1¼ inches.   7 Feb 1901

A pattern of the above-mentioned web sling has been sealed to govern future manufacture.

It is specially woven in one length, each end being fitted with a brass catch or hook.

The sling is attached to the rifle by passing the ends through the swivels from the outside, and then fastening them on to the body of the sling by means of the brass attachments.

It will supersede the following slings, existing stock of which will be used up—

    Slings, carbine, buff, R.A. †
    Slings, rifle, black (LoC 4697)
    Slings, rifle, buff. †
    Slings, rifle, buff, cadets, Sandhurst. †
    Slings, rifle, buff, mounted infantry. (LoC 7358)
    Slings, rifle, buff, short. †

† *Not published in List of Changes.*

10444–Saddlery, universal–   31 Dec 1900
      Frog, sword, saddle, Cavalry, Line. (Mark IV).

A pattern of the above-mentioned frog (as shown in the accompanying drawing) has been sealed to govern future manufacture.

It is shorter in the body than the Mark III, described in LoC 6671, and has only one hole for the shoe case strap; the steadying strap is attached to the body by a brass loop; and is fitted with a stud to allow it to be passed round the upper strap of the V attachment without unbuckling the girth straps.

10444

10471–Scabbard, brown, leather,   L   14 Mar 1901
    sword-bayonet, pattern 1888.

    Nomenclature.

    The nomenclature of the "Scabbard, sword-bayonet, pattern 1888" (LoC 10191) has been altered to that shown above by transferring the words "brown leather" from the detail column to the designation.

10472–Lockets, union, brass, Guards, Irish.   L   8 Aug 1900
    (Mark I).   31 Dec 1900
    S. and R. and F. with regimental device.

    Lockets, union, gilt, Guards, Irish.   L
    (Mark I).
    W.O. and S.S., with regimental device.

    A pattern of the above-mentioned ornament has been sealed to govern manufacture.

    It is fitted with an enamelled leather shield and two small leather blocks, and also, at the back, with four copper eyes, by which it can be secured to the flap of the valise with a small leather thong.

10505–Box, ammunition, S.A., home   L   24 Nov 1899
    and special. (Mark XIV).

    Brass pin and copper wire loop for securing lid.

    With reference to LoC 9516– A brass pin and copper wire loop will, in future, be used for securing the lid of the above-mentioned box, instead of a galvanized iron pin and iron wire loop as hitherto.

10558—Lance, exercise. (Mark IV). L   3 May 1901
Bamboo stave.

With reference to LoC 9516— It has been decided that, for future manufacture, the length, balance, and weight of the above exercise lance will be as follows—

Length from face of pad plate . . . . 8 ft. 7½ in. to 8 ft. 9½ in.
Balance from face of shoe-pad plate . . . . . . 3 ft. 7 in. to 3 ft. 10 in.
Weight of lance . . . . 3 lb. 4 oz. to 4 lb. 13 oz.

10559—Swords—   5 Mar 1901
    Cavalry, Household, pattern 1892   L
    Buglers   L
    Drummers (Mark I)   C
    Drummers (Mark II)   L
    Staff-Serjeants—
      Iron and gilt hilts   L
      Highland Regiments (Mark I)   L
      Highland Volunteers   L
      Pattern 1887   L

    Sword-bayonets—
      Lancaster carbine   L
      Pattern 1887   C
      M.H. rifle, converted   C
      M.H. Artillery carbine   L

Reduction in thickness of edges.
Sharpening before troops proceed on active service.

When the above-mentioned sword and sword-bayonets are passing through the Ordnance Factories for repair, the blades will have the edges reduced to a thickness of .01 inch as follows—

In the sword, Cavalry, Household, pattern 1892, the reduction of the front edge will commence about 8 inches from the hilt and of the back edge about 8 inches from the point, and continue to the point.

In the sword, buglers, and swords, drummers, the reduction of the edge will commence about 4 inches from the hilt on both edges and continue to the point.

In the swords, staff-serjeants, pattern 1889, iron and gilt hilts, the reduction of the front edge will commence about 8 inches from the hilt and of the back edge about 8 inches from the point, and continue to the point.

In the swords, staff-serjeants, pattern 1897, the reduction of the front edge will commence about 15 inches from the point, and of the back edge about 3 inches from the point, and continue to the point.

In the sword, staff-serjeants, Highland regiments (Mark I), and the sword, staff-serjeants, Highland Volunteers, the reduction of both edges will commence about 8 inches from the hilt, and continue to the point.

In the sword-bayonet, Lancaster carbine; sword-bayonet, pattern 1887; and the sword-bayonet, M.H. rifle, converted, the reduction of the front edge will commence about 6 inches from the hilt, and of the back edge about 6 inches from the point, and continue to the point.

Before troops proceed on active service, the above swords and sword-bayonets will be further sharpened, in accordance with the instructions contained in LoC 9206.

10560—Tool, clearing, .303-inch arms—   27 Mar 1901
    Bit, screw   C
    Screwed, to fit in bush.
    Bush, bit, screw   C
    Tapped, for bit and rod.

    Tool, clearing, .303-inch arms—
    Bit, corkscrew   C

Full size.

Bit, screw.

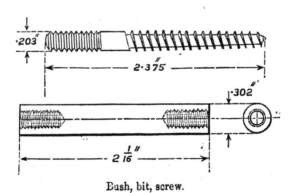

Bush, bit, screw.

To become obsolete.

Patterns of the above bit screw, and bush bit screw, have been sealed to govern future manufacture.

The bit is made of iron, and the bush of copper. They have been designed to take the place of the "Bit, corkscrew" (LoC 7817) and are used with the rod, tool, clearing, armourers, for removing material which may have become jammed in the bore of the barrels of .303-inch arms.

The "Bit, corkscrew," mentioned in LoC 7817, will accordingly become obsolete as those in Service wear out.

10561—Belt, pouch, buff, Cavalry, Line—   L   18 Dec 1900
Leather   15 Mar 1901
S.S., S.S. (Mark II), R and F., R. and F. (Mark II).
With buckle, slide, tip, and two studs.

> Pouches, ammunition, black, japanned, Cavalry, Line—
> Leather.
> **Carbine.** 20 rounds, S. & R. & F.
> **Carbine, .303-inch—**
> **(Mark I).** 30 rounds, S. & R. & F.
> **(Mark II).** Converted, 30 rounds, S. & R. & F.
> **(Mark III).** 30 rounds, S. & R. & F.
> **Pistol—**
> **S.S. and Trumpeters.** 24 rounds, fitted with tin.
>
> Pouches, ammunition, buff—
> **Expense, carbine—**
> **(Mark I).** Dragoons and Lancers.
> **(Mark II).** Cavalry.
> **Pistol, farriers, Cavalry, Line.** 12 rounds.
>
> Ornaments, brass, pouch—
> **1st Dragoons & 2nd Dragoon Guards.** Crown.
> **2nd Dragoons.** Eagle.
> **3rd Dragoon Guards.** Plume.
> **5th Dragoon Guards.** Star.
>
> Ornament, white metal, pouch—
> **6th Dragoons.** Castle.

Obsolete.

It having been decided that the pouch belts and pouches worn by Cavalry regiments of the Line are to be superseded by a bandolier for men armed with carbines, the above-mentioned articles are declared obsolete.

Until a satisfactory bandolier for revolver ammunition has been decided upon, the "Pouch, ammunition, brown, pistol, Infantry, Mark II" (LoC 5847) will be issued to men carrying revolvers, and will be worn on the belt, waist, brown, pistol, cavalry.

10637—Implement, action, pattern "D"   C   10 Jun 1901
All .303-inch magazine rifles and carbines.
With pouch for naval service only.

A pattern of the above has been sealed to govern future manufacture and conversion of "Implements, action, pattern 'C'."

It differs from the "Implement, action, pattern 'C' "(LoC 6981) only in the increase in the depth of the grooves for gauging the length of the striker point, i.e., from .037 and .041 inch to .040 and .042 inch.

"Implements, action, Pattern 'C'," in charge of the Naval, and Army, Ordnance Officers, will be sent to the Superintendent, Royal Small Arms Factory, Enfield Lock, for conversion to pattern "D". After conversion they will be returned to Ordnance Officers, who will exchange them for the "Implements, action, pattern 'C'," which are in the Service, and send the latter for conversion as above.

10638—Tools, armourers—   10 Jun 1901
    Gauges, armourers, striker point   C   18 Mar 1901
    pattern "C".

With reference to LoC 9040— The gauges, armourers, striker point, pattern "C", mentioned therein, will now be issued for land service.

All gauges, striker point, pattern "B" will be sent through the Naval, and Army, Ordnance Departments to the Superintendent, Royal Small Arms Factory, Enfield Lock, for conversion to pattern "C".

10639—Belts, waist, black—   L   17 Apr 1901
    Chief warders. Morocco.
    Warders. With snake hook.

Obsolete.

The substitution of brown for black leather in the equipment of chief warders and warders of military prisons having been approved, the above black waistbelts are declared obsolete, and will be dealt with accordingly, the following belts (LoC 4855) being issued in lieu—
    For chief warders—
        Belts, waist, brown, V.E., pattern 1882. Staff-serjeants.

For warders—
Belts, waist, brown, V.E., pattern 1882. Serjeants and rank and file.

Consequent upon the above, the black sword knot for chief warders will be withdrawn and used up for other services, the "knot, sword, brown, M.M.P., staff-serjeants," being issued in lieu.

The "pouches, warders," will be withdrawn and not replaced.

10640—Valise equipment, pattern 1888—   27 Jun 1901
   Belt, waist, buff, G.S. (Mark III).   L   10 Jul 1901
   With universal locket, S. and R. and F.

To be issued with the "Valise Equipment, pattern 1882."

When the store of "Belts, waist, buff, G.S. (Mark V), V.E., pattern 1882" (LoC 6504) is exhausted, the belt referred to above (LoC 5696) will be issued in lieu for use with the pattern 1882 equipment.

The rear end of the braces should be passed through the 1-inch buckles instead of through the 1-inch loops.

10687—Carbines, Artillery—   8 Dec 1900
   Martini-Enfield   L   19 Apr 1901
   Martini-Metford   L   20 Apr 1901

   Carbines, Cavalry—
    Magazine Lee-Enfield   L
    Magazine Lee-Metford   L
    Martini-Enfield   L
    Martini-Metford   L

   Rifles—
    Magazine Lee-Enfield   C
    Magazine Lee-Metford   C
    Martini-Enfield   L

Freedom of movement in nose-cap and fore-end.

It having been found that the shooting of rifles and carbines of the above-mentioned description was improved by permitting a slight movement between the barrel and fore-end, it has been decided that for future manufacture, and repair at Enfield and Birmingham, arms will have the barrel hole in nose-caps and upper bands, enlarged .02-inch, so as to give a clearance of .01-inch round the barrels. The barrel groove will also be enlarged, commencing at the lower band in rifles, and at the backsight in carbines, and increasing until flush with the barrel hole in nose-caps and upper bands.

Arms so altered have been issued since June 1901.

**10688—Tools, armourers—
  Tool, sight line, .03 inch left.**   L   12 Sep 1901
  For rifles, M.L.E. (Marks I and I*) and
  rifles, M.L.M. (Marks II and II*) fitted
  with barrel having Enfield rifling, with
  foresights .05-inch left; for renewing lines
  on slide and cap.

A pattern of the above-mentioned tool has been sealed to govern manufacture and immediate supplies for land service.

Full size.

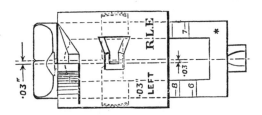

It is intended for use with rifle of the above-mentioned description that have foresights .05-inch left, and the sighting of which was corrected by replacing the sight leaves that had the V in cap and slide in the centre, with sight leaves with V in cap and slide .03-inch left (LoC 10303, 10439).

Officers having armourers tool boxes or chests on charge will put forward demands for one "Tool, sight line, .03-inch, left," per box or chest.

10746—Cartridges for Instruction.   C   4 Oct 1901

Precautions before drilling holes (LoC 10094).

In order to avoid the possibility of a "live" (Service) cartridge being dealt with as described in LoC 10094, all existing cartridges for instruction will, before having any holes bored in them, be carefully examined, and any about which there is any doubt (either from the label having become detached, or from any other cause), will be first tested by firing in a Service arm or machine gun of suitable calibre.

(Such cartridges can be fired into an earth butt or out to sea.)

10749—Box, ammunition, S.A. (Mark XV).   L   5 Apr 1900
Wood; with tin lining; 1,100 rounds, .303-inch.

A certain number of boxes of the above-mentioned description have been issued for land service, and a drawing (R.C.D. 9126A) has been sealed for record.

The sides and bottom of the box are of white deal, the ends of elm, the sliding top is of yellow deal, while the ends of the top are of Kawrie pine; the lining is of sheet tin.

The ends and sides are dovetailed together, and have the bottom and ends of the top secured to them by iron tinned screws. The ends of the top and sliding lid are tongued; the lid being secured in the closed position by a half round brass pin having a copper wire attached by which it can be removed from the box.

A rope grummet handle is spliced into two holes connected by a groove in each end of the box.

Dimensions, &c.
Exterior overall for stowage, not to exceed—
Length . . . . .22.187 ins.
Width . . . . 8.5 ins.
Depth . . . . 7.02 ins.
Tonnage . . . . .02 ton

10750—Box, ammunition, S.A.,   L   3 Nov 1898
750 rounds .303-inch. (Mark I).   30 Nov 1898
Teak; with tin lining.   22 Jun 1900

A pattern of the above-mentioned box has been sealed to govern manufacture for land service as may be required.

It is made of teak, and is provided with a tin lining (which has a closing plate fitted with a handle) and a sliding lid secured by a brass pin and copper wire loop. A cleat with rope handle for lifting is secured to one end of the box.

Dimensions, &c.
Length, overall .13.875 ins.
Width . . . . 8 ins.
Depth . . . . 8 ins.
Weight. . . . . 9 lb. 5 oz.

10751—Box, ammunition, small-arm, pistol,   L   2 Apr 1900
Enfield, 240 rounds. (Mark III).
Wood, with tin lining, also 276 Mark II, and 300 Mark III, ball, Webley, and 420 rounds, blank, Webley.

Alteration to closing plate.

In future manufacture, the closing plate of the above-mentioned box (LoC 4094, 7488) will be altered so as to admit of a larger opening in the tin lining when the closing plate is removed.

Consequent upon this alteration, the numeral of the box will be advanced to "Mark III" as shown above.

10763—Guards, hand—   27 Aug 1901
    Wood   1 Oct 1901
      M.L.M. rifle Mark I* and Mark II   C
      and II*; also M.L.E. rifles.

      M.E. rifle Mark I & Mark II   L

Repair of.

In future, handguards of the above-mentioned description which have longitudinal splits will be repaired locally as described below, by the insertion of two steel wire rivets, No. 16 W.G., as shown in the accompanying drawing.

Such repair to handguards in possession of Regular and Militia units will be carried out by regimental and circuit armourers, who will be allowed 1 yard of steel wire annually per 1,000 arms. This wire will be obtained from Army Ordnance Department on demand in the usual way.

Handguards split in course of manufacture will be similarly repaired at the Royal Small Arms Factories. Such handguards, however, will not be issued from the factories with new arms, but will be utilized for repair of arms and for issue to the troops, &c.

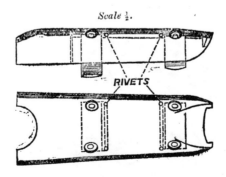

Instructions for repair.

A little glue should be applied to the split in the handguard, which should be placed on the rifle or mandril of the same size; the holes should then be drilled for the rivets, and the rivets driven in and cleaned off. The handguard should be allowed to remain in the rifle or mandril until the glue is thoroughly dry.

10764—**Bandolier, leather, .303-inch ammun-**   L   19 Jun 1901
    ition, 60 rounds. (Mark I).
    With steadying strap.

    Bandolier, leather, .303-inch ammun-   L
    ition, 50 rounds.
    With steadying strap.

Nomenclature.

A pattern of the above-mentioned bandolier, as shown in the accompanying drawing, has been sealed to govern future manufacture for use by Cavalry and Mounted Infantry. It will supersede the "Bandolier, leather, .303 ammunition" (LoC 5873, 8789) from which it differs in being lighter, the body narrower and fitted with 12 pockets riveted to it, each pocket holding five cartridges.

The divisions of the cartridges are formed by a leather lace passing through the pocket from back to front and the ends riveted; the nose of the bullets project through the holes in the bottom of the pocket.

The weight of the bandolier is 1 lb. 6 oz.

Consequent on the introduction of this bandolier, the nomenclature of the pattern referred to in LoC 5873, 8789, has been amended as shown above. No more of the 50 rounds bandolier will be made, but existing stock will be used up.

10765—**Sabretaches**— 7 Nov 1901
    Without carriages.
    Staff serjeants   L
    Black japanned leather.
    Other ranks   L
    Black leather.

    Carriages, sabretache, with billets—
    Staff serjeants   L
    Buff, with gilt buckle.
    Other ranks   L
    Buff, with brass buckle.

1. Obsolete.

    Belts, waist, buff, Cavalry, pattern 1885—
    Staff serjeants   L
    Other ranks   L

2. Removal of chape and dee at the back.

    1. It having been decided to abolish the sabretache throughout the service, the above-mentioned articles are declared obsolete, and will be returned to store for disposal, together with all material special for their repair.

    2. In consequence of this decision, the above-mentioned belts (LoC 4853) will, in future manufacture, be modified by the omission of the chape and dee (intended to take one of the straps that support the sabretache) at the back of the belt.

    Belts in store and in the hands of troops will be altered locally by cutting the stitches connecting the chapes to the belt and removing the chape and dee, which will be returned to store for disposal.

10797—**Chest, carbine, with fittings D**   L   10 Nov 1900
    Case, regimental, 10 carbines, M.L.M.,   8 Jan 1901
        cavalry, pattern A.   L

    Bridge battens strengthened by wire nails.

    In future manufacture, the bridge battens of "Chests, carbine" (LoC 8612) and "Cases, regimental" (LoC 7878), of the above-mentioned description will be strengthened by steel grooved wire nails.

10798—Rifles, magazine, Lee-Metford,   C   21 Sep 1901
       Marks II and II*.   24 Oct 1901

       Rifles, magazine, Lee-Enfield,   C
       Marks I and I*.

Correction of sighting.

With reference to LoC 10393, 10439— Any rifles of the above-mentioned description in the hands of troops, that have the V-notches in the back sight .03-inch left, and shoot 5 inches or more left at 200 yards, may be fitted with a leaf having the V notches in cap and slide central.

Leaves for this purpose will be demanded from Army Ordnance Department as required.

10844—Boxes, ammunition, S.A.—   7 Aug 1901
    .303-inch, half, naval   N   12 Dec 1901
    G.S.   C
    Home & special—
      Mark XII   L
      Mark XIV   L
      Mark XV   L

Linings.

In future manufacture, the linings of ammunition boxes of the above-mentioned descriptions will have the end seams capped and cramped, instead of knocked up as hitherto.

10974—Swords—   19 Dec 1901
    Buglers (Mark II)   L
    Cavalry, Household, pattern 1892
      (Mark II)   L
    Cavalry, pattern 1899   L
    Drummers (Mark II)   L
    Mountain Artillery   L
    Naval—
      Pattern 1900   N
      Pattern 1889   N
      27-inch (Mark II)   N
    Pioneers   L
    Staff-Serjeants, Highland Regiments   L
    Staff-Serjeants, pattern 1898   L

Buff pieces.

To be reduced in size and waterproofed in future manufacture.

In future manufacture, in order to prevent water getting into the scabbard, the buff pieces will be reduced to the size of the mouthpieces of the sword scabbards, and will be waterproofed by being soaked in melted paraffin wax.

11033—Carbines, .303-inch.
   Rifles, .303-inch.

Gauging and exchange of barrels when bore is worn.

It has been decided, in future, to exchange the barrels of all .303-inch rifles and carbines when the bore is so worn as to take the following gauge plugs—

1. The .308-inch plug— ¼ inch at muzzle.
2. The .309-inch plug— ¼ inch at breech.
3. The lead plug— ¾ inch at breech.
4. The lead plug— ¼ inch at breech, in conjunction with the .308-inch plug— ¼ inch at breech.

In land service these plugs will be used by the viewers of the Small-Arms Inspection Department in their annual inspection of the arms at stations at home.

They will also be issued to Chief Ordnance Officers at stations abroad for use when examining arms.

In naval service these plugs will be used at Naval Ordnance Depots at home and abroad, and will be supplied on receipt of demands.

11034—Guards, hand—   17 Feb 1902
   Wood.
   M.L.M. rifle (Mark II)   C
   also Mark II* and M.L.E. rifles.

   M.E. rifle (Mark I) & (Mark II)   L

Repair of.

With reference to LoC 10763— The drawing therein, showing the position of the rivets in the handguard, is applicable to the handguard of the M.L.M., Mark I* only.

In handguards of rifles, M.L.E.; M.L.M., Marks II and II*, and M.E., Marks I and II, the positions of the rivets should be as shown in the accompanying drawings respectively.

The allowance of steel wire, No. 16, W.G., for the repair of all handguards is increased to 3 ounces per 1,000 arms.

Scale ½.
M.L.M. (Marks II., II.*) and M.L.E. rifles.

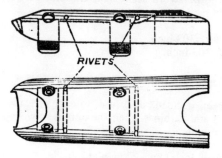

Rifle, Martini-Enfield (Mark I.).

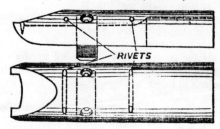

Rifle, Martini-Enfield (Mark II.).

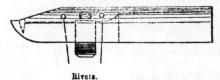

Rivets.

The drills required for boring the holes, for the rivets, in the handguards should be made by the armourer from the wire supplied for the rivets.

11035—Tools, armourers—   10 Oct 1901
    Wrench, removing screws, bolt-head,
    rifle, magazine, Lee-Metford, Mark I*,
    (Mark I).   L
      With three spare screw drivers.

    A pattern of the above wrench has been sealed to govern manufacture.

    The wrench is designed for the removal of the screws of bolt-heads in the above rifles that have become too tight for removal with the "armourers' brace" and "bit, screwdriver."

It should be used in the following manner—

Push back the wedge, raise the screw sufficiently to enable the bolt-head to be inserted on the seating, turn down the screw head until the screw-head is exposed, then turn down the screw of the tool until the driver nearly touches the screw of the bolt-head, and is in line with the slit. Press in the wedge until the screwdriver touches the bottom of the slit in head of screw, and unscrew with handle of tool.

To take out the bolt, push back the wedge, and raise the screw of tool.

Two of the tools, each with three spare screwdrivers, will be issued to the Chief Ordnance Officer of each district. They may be obtained by armourer-serjeants on loan when required.

11037—Pouches, ammunition, black—			11 Oct 1900
Artillery. (Mark II).		L	5 Nov 1901
Leather; 20 rounds with gun ornament, limber gunners.			
Cadets, Sandhurst. (Mark II)		L	
Morocco leather; 20 rounds.			
Cadets, Woolwich. (Mark II)		L	
Morocco leather; 20 rounds, with gun ornament.			
Pouches, writing materials—			
Conductors. (Mark II).		L	
Morocco leather, with E.R. ornament.			
Engineers. (Mark II).		L	
Morocco leather; W.O. and S.S., with Royal Arms ornament.			

Patterns of the above-mentioned pouches have been sealed to govern future manufacture.

They are similar in shape to the previous patterns (Mark I) described in LoC 3430, 3110, 3748 and 2331, but with the exception of the "pouch, ammunition, artillery," are made of Morocco leather, instead of Japanned leather.

The writing pouch is made of plain black leather.

The "pouch, writing materials, conductors," will guide manufacture for the "pouch, writing materials, R.A." the latter differing only from it in having a gun ornament instead of an E.R. ornament.

11078—Carbine, magazine, Lee-Metford. (Mark I) L   13 Jan 1902

    Fitting of Lee-Enfield barrel.
    Alteration of nomenclature.

With reference to LoC 7751, 8248 and 8676— When carbines of the above description are fitted with Lee-Enfield barrels, and have the wings of the nosecaps drawn out to the same height as those on Lee-Enfield carbines, they will be described as—
    Carbines, magazine, Lee-Enfield. (Mark I).   L

    The barrels will be marked by the manufacturer on the knox-form with the letter "E".

Instructions regarding local conversion.
When barrels of M.L.M. carbines require replacement, and are demanded from Army Ordnance Department, Ordnance Officers will, if barrels with Metford rifling are not available, issue in lieu, "Barrels with bodies, M.L.E. carbine," and a corresponding number of "Caps, nose, M.L.E. carbine." The latter will be fitted to the carbines at the same time as the new barrels, and the surplus M.L.M. nosecaps will be returned to store for transmission to Birmingham for conversion. These conversions will be recorded on Army Form G 1049, which will be posted in the Equipment Account.

11079—Tools and materials, armourers—   14 Jun 1901
    Rod, cleaning—   2 Oct. 1901
      No. 1, .303-inch   C   9 Jan 1902
      With swivel handle, for use with jute.   15 Feb 1902
      Rifle and carbine.   30 Apr 1902
      No. 2, .303-inch   C
      With swivel handle, for use with brass wire.
      Rifle.
      No. 3, .303-inch   L
      With swivel handle, for use with brass wire.
      Carbine.

    Muzzle guides—
      Rifle and carbine, Artillery, .303-inch.   C
      For use with cleaning rods.
      Carbine, Cavalry, .303-inch.   L
      For use with cleaning rods.
    Wire, brass, No. 26, W.G., hard,
      3-inch lengths.   C
      For use with cleaning rods, Nos. 2 and 3.
    Jute, dressed   C
      For use with rod, cleaning, No. 1.
    Tools, clearing, .303-inch arms—
      Rod, No. 2   C
      With strengthened handle, for all .303-inch arms,

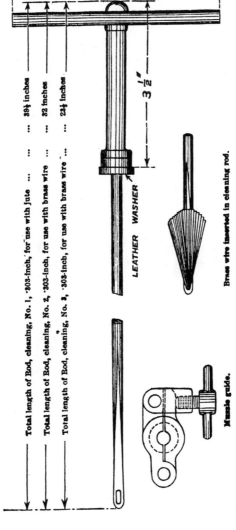

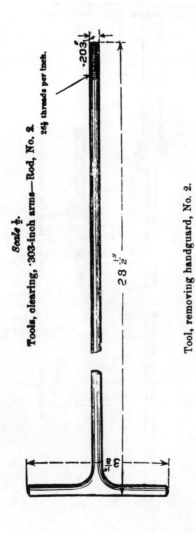

Tools, clearing, .303-inch arms—Rod, No. 2.

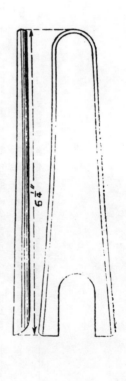

Tool, removing handguard, No. 2.

screwed at end to take bit, screw; bush, bit, screw; bit, spoon; and plug, clearing, plain.
Tool, removing handguard, No. 2   C
Rifles, magazine and M.E.

1. New patterns introduced.

Tools, armourers'—
Rod, cleaning   C
End tapped to receive wire brush.
Brush, wire   C
Bit, corkscrew   C
All .303-inch rifles and carbines, tapped to fit on rod.
Rod, tool, clearing   C
All 303-inch rifles and carbines; screwed at end to take bits; also for plug, clearing, plain.

2. Patterns to become obsolete.

The following changes in patterns of tools and material for arms have been approved.

1. Stores of new pattern introduced.

Rods, cleaning—
    No. 1, .303-inch. For use with jute. To enable armourers to clean barrels that are rusty in bore. The rod for jute is made in one length, suitable for carbines and rifles. It is to be used with jute and flour emery.
    No. 2, .303-inch. For use with brass wire; rifle. To enable armourers to clean barrels that are rusty in the bore. The rods for brass wire are made in two lengths, as the wire should not pass beyond the front end of chamber.
    No. 3, .303-inch. For use with brass wire; carbine. To enable armourers to clean barrels that are rusty in the bore; as No. 2.

Muzzle guide—
    Rifle and carbine artillery, .303-inch; and Carbine, cavalry, .303-inch. For use with cleaning rods. This is fixed on the muzzle to prevent the rod from rubbing the bore at the muzzle. To use the guide with carbines it will be necessary to remove the nosecap.

Wire, brass, No. 26, W.G., hard, 3-inch lengths. For use with cleaning rods. This will be issued from store, cut in 3-inch lengths; 60 strands should be inserted in the eye of the rod and bent backwards towards the handle.

Jute, dressed. For cleaning the bore of barrels. This should be cut by armourers into 8-inch lengths. It will be issued from store. For land service— in hanks of about 1 lb. each.

For naval service— in accordance with arrangements that will be made by the Admiralty.

Tools, clearing, .303-inch arms—
Rod, No. 2. For use with bits, screw; bush, bit, screw; bit, spoon; and plug, clearing, plain. The rod is made with solid T handle to strengthen it.

Tool, removing handguard, No. 2. For removing handguard on rifles, L.M., L.E., and M.E. This is applied as a lever under the front end instead of as a wedge at the back, as in the case of the former pattern.

2. Tools, armourers, to become obsolete.

LoC 6379— Rod cleaning— To be superseded at once by rods, cleaning, Nos. 1 and 2 for rifles, or Nos. 1 and 3 for carbines.

LoC 6379 & 7324— Brush, wire, M.L.M. rifle; also all .303-inch carbines and rifles. To be superseded at once by dressed jute and brass wire.

LoC 7817— Rod, tool clearing, all .303-inch rifles and carbines, screwed at end to take bits, also for plug, clearing, plain. To be superseded by rod, tool, clearing, No. 2, when present stock is used up and those in use become unserviceable.

LoC 6379— Tool, removing handguard. To be superseded by tool, removing handguard, No. 2, when present stock is used up, and those in use become unserviceable.

Officers concerned will at once take steps as follows—

Land Service:

(a) Put forward demands for stores of new pattern in the following proportions—
Rods, cleaning—
No. 1: 1 for each armourer.
No. 2: 1 for each armourer, according to arm in possession.
No. 3: 1 for each armourer, according to arm in possession.
Wire, brass, No. 26, W.G., hard, 3-inch lengths: 2 oz. per 100 arms.
Jute, dressed: 1 hank for every 200 arms in possession.
Muzzle guide—
Rifles and carbines, artillery: 1 for each armourer, according to arm in possession.
Carbine, cavalry: 1 for each armourer, according to arm in possession.

(b) On receipt of above, return the obsolete articles to store for disposal.

Naval Service:

Supplies will be made in accordance with instructions, which have been issued by the Admiralty.

11080—Ornament, gilt, pouch, Conductors. 5 Nov 1901
  (Mark II).
  and W.O., R.A.M.C.; E.R. and crown.

A pattern of the above-mentioned ornament has been sealed to govern future manufacture.

It differs from that referred to in LoC 3748 in the monogram and design of the crown.

11109—Scabbard, brown leather, sword- N 23 Sep 1901
  bayonet, pattern 1888, naval. (Mark I) 7 Feb 1902
  10 Apr 1092

A pattern of the above-mentioned scabbard, a drawing of which is attached, has been sealed to govern future manufacture.

It differs from the "Scabbard, sword-bayonet, pattern 1888, Mark I" (LoC 5877), in being made of brown leather, with leather tip instead of the steel chape, and the stud on the locket being chequered on top.

Scale ¼.

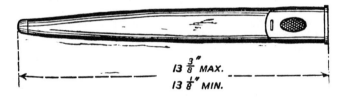

11110—Accoutrements, naval, pattern 1901— 8 Aug 1901
  Bag, ammunition, (Mark I) N 28 Feb 1902
  Brown leather.
  Bandolier, naval. (Mark I) N
  Brown leather, 60 rounds with
    steadying strap.
  Belt, waist. (Mark I) N
  Brown leather.
  Braces, pairs. (Mark I) N
  Brown leather.
  Carriage, waterbottle. (Mark I) N
  Brown leather, with shoulder strap.
  Case, pistol. (Mark I) N
  Brown leather.
  Cover, mess tin. (Mark I) N
  Brown canvas, with 4 leather loops.

Frog. (Mark I).   N
Brown leather, with shifting loop,
 sword-bayonet or cutlass.
Haversack. (Mark I).   N
Brown duck, with removeable shoulder
 strap.
Mess tin. (Mark I)   N
Tin; with brass wire handle.
Pouch, cartridge, pistol. (Mark I).   N
Brown leather, 42 rounds.
Strap, mess tin. (Mark I).   N
Brown leather, 25 inches by ¾ inch.

Patterns of the above-mentioned articles have been sealed to govern manufacture.

They differ from all previous patterns of naval accoutrements in the leather employed, which is of a darker shade of colour, and much thinner. They will only be supplied as the stock of existing pattern becomes exhausted.

The "Bag, ammunition," differs from the previous pattern, (LoC 7576) in being about one-third deeper. This extra depth is made of flexible leather, and can be turned up when empty.

The "Bandolier" differs from that shown in LoC 10764 by the pockets being deeper and tapered to take five cartridges in a charger, and the leather lace is dispensed with.

The "Belt, waist," consists of a plain strap, similar to the "Belt, waist, for cutlass" (LoC 3969, 5138).

The "Braces, pairs," differ from the previous pattern (LoC 8362) in having both ends fitted with hooks, which, in front, take the place of rings and buckles. The braces cross at the back through a loop.

The "Carriage, water-bottle," is similar in form to that shown in LoC 4974, but a ring is fitted on each side at the top of the carriage portion, to allow the shoulder strap to pass through freely. A metal clip is also provided to pass behind the waist belt for steadying purposes. (The water-bottle is of a new pattern, and particulars regarding it will shortly be published).

The "Case, pistol," differs from the previous pattern (LoC 6434) in being made a little longer, and in the loops on the back being placed in a different position. It is similar in shape to that shown in LoC 10440.

The "Cover, mess tin," is made of sail canvas (dyed brown),

fitted with a brass stud, one billet, and four leather loops. The cap front is bound with leather.

The "Frog" differs from the previous pattern (LoC 6434) in being smaller. The front is blocked up to enable it to take the scabbard, and it is provided with a shifting loop and stud.

The "Haversack" is made of linen duck (dyed brown), and is fitted with a partition, a 2-inch wide shoulder strap 50 inches in length, and a fixed strap 10½ inches in length and 1¼ inches in width. The haversack is of the following dimensions, viz.—

	Front	Back
Depth (including flap)	11¼ in.	17½ in.
Width (at top of bag)	10¼ in.	10¾ in.
Width (at widest part)	12 in.	12 in.

The "Mess tin" is D-shaped, and consists of body, lid, and tray. The lid is fitted with a tinned-iron wire handle, so that it can be used as a frying pan, and the tray fits into the upper part of the body.

The "Pouch, cartridge, pistol," is a small open pouch to carry ammunition in packets; six pockets (each to take one cartridge) are fitted on the outside of the front of the flap.

The "Strap, mess tin," is a plain strap with a small brass loop fitted about 9 inches from the buckle end.

Carriages, blanket, are not required with the above accoutrements.

11111—Accoutrements, naval, pattern 1901   8 Aug 1901
    Becket, sword. (Mark I).   N
    Brown leather.
    Sling, rifle. (Mark I).   N
    Brown leather, 39 inches by 1 3/8 inches, with stud.

Nomenclature.

The nomenclature of the rifle sling referred to in LoC 7188, and of the becket referred to in LoC 3969, has been altered to read as above.

11112—Ornament, gilt, pouch, Engineers.   28 Apr 1902
    (Mark II).   L
    W.O. and S.S.; Royal Arms, and regimental motto.

A pattern of the above-mentioned ornament has been sealed

to govern future manufacture.

It differs from the previous pattern worn on the pouch mentioned in LoC 2231, in general design of details.

11113—Plates, waist-belt—   11 Jan 1902
    Brass, Field Artillery. (Mark II).   L   21 Apr 1902
    With ornament, drivers and
    dismounted men.
    Gilt, Artillery. (Mark II).   L
    With ornament, with catch,
    W.O. and S.S., also R.E.

Patterns of the above-mentioned plates have been sealed to govern future manufacture.

They differ from the previous patterns in the design of the crown.

11150—Case, 200 sword-bayonets, pattern 1888   C   28 Nov 1901
    Chest, carbine, pattern A   L
    Chest, carbine, pattern B   L
    Chest, carbine, pattern C   L
    Chest, rifle, pattern A   C
    Chest, rifle, pattern B   C
    Chest, rifle, pattern C   C
    Chest, rifle, pattern D   C
    Fitted with battens.

Existing cases and chests of the above-mentioned descriptions when returned to Weedon will, when found to require it, have the ends strengthened by the addition of battens.

For future manufacture, "Cases, 200 sword-bayonets, pattern 1888". (LoC 6146), "Chests, carbine, pattern C" (LoC 7268), and "Chests, rifle, pattern C" (LoC 7269), will have the ends strengthened in a similar manner.

11151—Sword-bayonet, pattern 1888. (Mark III)   C   23 Sep 1901
    All .303-inch rifles and carbines fitted   17 Dec 1901
    to take the pattern 1888 sword-bayonet.

    Scabbard, brown leather, sword-bayonet,   L
    pattern 1888, land. (Mark II).
    Except for the Foot Guards.

Patterns of the above have been sealed to govern future manufacture.

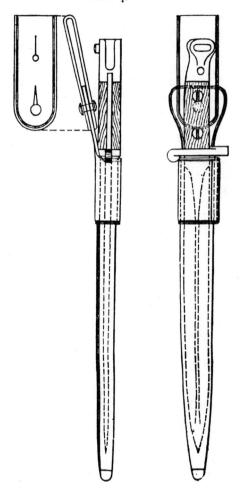

The sword-bayonet only differs from the sword-bayonet pattern 1888, Marks I and II, described in LoC 5877 and 9701, in the tang, pommel, and crosspiece, which are browned, and in the grips, which are fixed with two screws and nuts (oil blacked), to allow of the grips being removed when the bayonet requires re-browning.

The scabbard differs from the scabbard, sword-bayonet, pattern 1888, Mark I, described in LoC 5877, in being made of brown leather with a leather tip instead of a steel chape.

The locket is covered with a band of brown leather which is extended upwards to form a tag.

A brass double button is fitted to the tag for attaching it to the loop, scabbard (LoC 11154), by means of which the scabbard is suspended on the waist belt. This pattern of scabbard cannot be worn with the frog on.

Sword-bayonets, pattern 1888, Marks I, II, and III are interchangeable in scabbards, sword-bayonet, pattern 1888, Marks I and II. Some of the earliest scabbards of this pattern have the stitching of the leather band on the opposite side of the scabbard to that shown in the drawing.

11154—Loops, scabbard—   14 Jan 1902
    Leather, 9½ inches by 1¾ inches;
    scabbard, brown leather, sword-bayonet,
    pattern 1888, land, Mark II.
    Black (Mark I).   L
    Brown (Mark I).   L
    Buff (Mark I).   L

Patterns of the above-mentioned loops have been sealed to govern manufacture.

They are 9½ inches in length by 1¾ inches in width, and are provided with three button holes (see illustration of the loop in the drawing depicting the Mark II scabbard, LoC 11151). The centre one is intended to be used when the scabbard is carried on the brass loop attached to the web waist belt, the other two are to be secured to the stud on the scabbard when the latter is carried on an ordinary leather waistbelt.

11202—Accoutrements, naval, pattern 1901—   25 Jun 1902
    Case, pistol (Mark II).   N
    Brown leather.

A pattern of the above-mentioned case has been sealed to govern future manufacture, and the alteration of existing store of Mark I cases.

It differs from the previous pattern (LoC 11110) in the muzzle portion being 1¼ inches shorter, and the position of the upper back loop being slightly altered.

The front of the case has also been reduced to admit of the revolver being more easily withdrawn.

Mark I cases should be altered locally; an altered case to guide the alteration will be supplied from Naval Ordnance Store, Woolwich, to the depots concerned, in accordance with arrangements which have been made by the Superintendent of Ordnance Stores.

11203 – Ornaments –   2 Jun 1902
    Brass, plate, waistbelt, Household
      Cavalry (Mark II).   L
    All ranks, also pouch, Mark III,
      R. & F.; Royal Arms.
    Gilt, pouch, Household Cavalry
      (Mark II).   L
    Staff and band; Royal Arms.

Patterns of the above-mentioned ornaments have been sealed to govern future manufacture.

They differ from those referred to in LoC 4161 and 4490 in the design of the crown.

11208 – Bucket, rifle. (Mark IV).   9 Jun 1902
    Leather, with 1 steadying and
    2 suspending straps.

A pattern of the above-mentioned rifle bucket has been sealed to govern future manufacture.

It differs from Mark III (LoC 9103) in being 2¼ inches shallower in the body, and in being shaped to allow the toe of the rifle butt to be carried front or rear.

A 1-inch brass link is fitted at each end of the body, and a detachable steadying strap is furnished to allow the bucket to be worn on the near or off side of the horse.

The chapes for supporting straps are inserted between the lining and the outside of the lay of bucket. The buckles are brass instead of iron.

11266 – Bandolier, web, double loop, 100 rounds L   16 Jul 1902
        Bandolier, web, double loop, 70 rounds   L
        Bandolier, web, double loop, with
          cover, 70 rounds   L
        Bandolier, web, double loop, with
          cover, 50 rounds   L
        Belt, waist, web, 80 rounds   L
        Belt, waist, web, with covers, 80 rounds L
        Belt, waist, web, double loop, with
          covers, 60 rounds   L

With steadying strap.

Supplies of the above-mentioned bandoliers and belts have been issued to the troops. No more will be obtained, but existing stock will be used up.

The "Bandolier, web, double loop, with cover, 70 rounds," in store, and in use by the troops, will be fitted with an additional stud and tab for securing the cover, in accordance with the following instructions, the undermentioned articles being supplied from store for this purpose to officers concerned on demand—

    Studs, brass, bandolier, with washer . . . . 2 per bandolier.
    Tabs, brown, leather . . . . 1 per bandolier.

Instructions.

Remove the central cover tab and stud from the bandolier; fit on the two studs supplied equidistant from end studs (i.e. between 12th and 13th double loop from each end); sew leather tabs on cover in position corresponding to the two central studs. Studs and washers taken off the bandoliers to be returned to store.

11267—**Belts, shoulder, brown, sword,**
            12 Mar 1902
    "Sam Browne", pair. (Mark II)   L   16 Jul 1902
    Leather, with 2 billets.
    Belt, waist, brown, sword,
    "Sam Browne". (Mark II)   L
    Leather, with buckle, frog, and 21-inch
    by 5/8-inch strap.
    Case, brown, pistol, "Sam Browne",
    (Mark II)   L
    Leather, with cover.
    Pouch, ammunition, brown, pistol,
    "Sam Browne". (Mark II)   L
    Leather.

Patterns of the above-mentioned articles have been sealed to govern future manufacture.

They differ from the previous patterns LoC 10440 in being made of lighter materials throughout. The waistbelt also is narrower and the frog is smaller.

11289—**Carbines**—   2 Sep 1901
    M.L.E., Cavalry. (Marks I and I*)   L   21 Sep 1901
    M.L.M. (Mark I)   L   10 Mar 1902
    Rifles—   20 Mar 1902
    M.L.E. (Marks I and I*)   C   20 Apr 1901
    M.L.M. (Marks I*, II and II*)   C   14 May 1901

1. Treatment of stocks, butt, as a preventative against shrinkage.
2. Modifications to prevent stock bolt working loose.
3. Percentage of "long" and "short" stocks, butt, in future issues of the M.L.E., Mark I* rifle.
4. New components introduced.

1. With a view to preventing shrinkage and working loose of stocks, butt, the socket end of butts will, in future manufacture, be soaked in a preparation of benzole and paraffin wax, and then compressed to size.

Butts so treated will be distinguished by the letter "P" stamped on the right side near the socket end before issue. When butts so marked are attached by armourers, the "Anvil, stock, butt," will be used as described in LoC 11293.

2. Stock bolts of future Ordnance Factories manufacture will, to guard against any possibility of their working loose and becoming unscrewed, have the end squared to fit into a square recess in a keeper plate which is let into the rear end of the fore-end. Stock bolts of this pattern will be known as No. 2, and those in arms of the above description passing through the Royal Small-Arms Factory, Sparkbrook, for repair will be converted to this pattern.

In consequence of the modification to the stock bolt, the stock, butt, will have the large hole for the bolt deepened to allow of the squared end of the bolt projecting beyond the inside face of the socket of the body. Stocks, butt, taking the No. 2 pattern stock, bolt, and stocks, fore-end, fitted with the keeper plates, will be marked with a "2" on the right side near the socket of body, and will be known as No. 2.

3. In order that the soldier may have a rifle that fits him properly, in future issues of the M.L.E., Mark I*, rifle, 10 per cent. will have stocks, butt, ½ inch longer, and 10 per cent. ½ inch shorter † than the present length.

These will be recognised by the letters "L" or "S" marked ½ inch in front of the tang of the butt plate. Care should be taken to issue rifles fitted with above butts only to men that they suit.

4. Consequent on paras. 2 and 3 above, the following new components have been introduced into the Service—

Bolt, stock, No. 2, all magazine rifles and carbines	C
Plate, keeper, stock bolt, all magazine rifles and carbine	C
Stocks, butt, No. 2—	

Normal—
 M.L.M. rifle, Mark I*   C
 M.L.M. rifle, Mark II   C
Long—
 M.L.M. rifle, Mark II   C
Short—
 M.L.M. rifle, Mark II   C
 (and Mark II*, also M.L.E. rifles for the 3 lengths)
 M.L.M. carbine, Mark I   L
 and M.L.E. carbines.
Stocks, fore-end, No. 2—
 M.L.M. rifle, Mark I*   C
 M.L.M. rifle, Mark II   C
 and II*; also M.L.E. rifles.
 M.L.M. carbine, Mark I   L
 and M.L.E. carbines.

Note for NAVAL SERVICE— The above modifications, &c., will be embodied in rifles that may be manufactured in future for naval service, but the new pattern stocks will not be fitted to rifles returned to Royal Naval Ordnance Depots for repair until the old pattern stocks are used up.

† In order to get the pull-through into the smaller recess in the short butts, it must be very tightly wound, in accordance with the instructions given in "Care and Description of Arms and Ammunition" in the Musketry Regulations.

11290—Dirk (Mark II)   L   20 Aug 1902
 All pipers and band, and drummers of   18 Mar 1902
 kilted Highland regiments.   19 Mar 1902
 Sword, Cavalry, Household, pattern 1892,
  (Mark II)   L
 Sword, staff-serjeants, pattern 1898
  (Mark I)   L
 All dismounted services except regiments
 having claymores.

Sealing of new patterns.

Patterns of the above-mentioned have been sealed to guide future manufacture. The numerals have not been advanced.

The "Dirk, Mark II" differs from that described in LoC 3504 only in the form of crown on the cap of hilt.

The "Sword, Cavalry, Household, pattern 1892, Mark II" differs from the pattern described in LoC 8581 only in the form of crown on the hilt; and the "Sword, staff-serjeants, pattern 1898, Mark I" from that in LoC 9243 in the form of the crown and the substitution of "E.R." for "V.R." in the cypher on the hilt.

11291—Drivers, screw, armourers, medium.   L   16 Sep 1902

To become obsolete.

The "Driver, screw, armourers, medium," will become obsolete when the present stock is exhausted and those in use are worn out.

The "Driver, screw, armourers, large," will then be issued and used for the purposes for which the medium pattern is now employed.

11292—Sword-bayonets, pattern 1888.   4 Sep 1902
(Marks I, II & III).

Marking.

In future, sword-bayonets of the above description having blemishes which cannot be removed from the pommel, but which do not affect the serviceability of the bayonets, will be marked with a star on the end of the pommel. This star will only be added at the Ordnance Factories as the sword-bayonets pass through for repair.

11293—Tools, armourers—   28 Nov 1901
    Anvil, stock, butt, magazine rifles   10 Jul 1902
    and carbines.   C
    Gunmetal; for attaching stocks, butt.

A pattern of the above-mentioned anvil has been approved for the use of armourers when attaching stocks, butt, which have been treated with paraffin wax and benzole, and compressed (LoC 11289).

The anvil is provided with two studs to retain it in position when on the butt— one stud to enter the stock-bolt hole, and the other the butt-plate screw hole. The butt will be inserted in the socket of the body, the anvil placed in position on the end of the butt, which will then be driven home into the socket with either a mallet or a hammer. The anvil should then be withdrawn, and the stock bolt inserted and screwed up.

For Land Service, officers concerned should demand one anvil per armourer serjeant under their orders.

For Naval Service, this tool will not be required at present (vide LoC 11289).

11294—Tools, armourers—   12 May 1902
    Driver, screw, forked   C   3 Sep 1902
    For nut, screw grip, sword-bayonet, pattern 1888. (Mark III).

A pattern of the above-mentioned screw-driver has been sealed to govern manufacture. It is to be used to hold the nut of the grip screw while the screw is being tightened up by a screw-driver.

11295—Tools, armourers—   28 Jul 1902
    Mandril—
        Scabbards, brown leather, sword-
        bayonet, pattern 1888, Land,
        Mark II and Naval Mark I—

        Blocking leather   C
        Gunmetal.

A pattern of the above-mentioned mandril has been sealed to govern manufacture.

This mandril is special to the above-mentioned scabbards.

11296—Lockets, union, gilt—   1 Sep 1902
    Royal Marines. (Mark III)
    W.O. and S.S., lion, crown, and regimental motto.
    Universal. (Mark III)
    W.O. and S.S., lion, crown, and royal motto.
    Lockets, union, brass—
    Royal Marines. (Mark III)
    S. and R. and F., lion, crown, and regimental motto.
    Universal. (Mark III)
    S. and R. and F., lion, crown, and royal motto.

Patterns of the above-mentioned lockets have been sealed to govern future manufacture.

They differ from those referred to in LoC 10441 in the design of the crown.

11316—Box, ammunition, small-arm—   7 Mar 1902
    G.S., Land (Mark XI)   L
    G.S., Naval (Mark XI)   N

    Dimensions.

In future, the internal dimensions of Mark XI small-arm ammunition boxes (LoC 4021) will be assimilated to those of the Mark XIV box (LoC 6588), the dimensions of the latter being the most suitable for the linings.

The dimensions, &c., of the Mark XI box as altered, are as follows—

	Length.	Width.	Depth.	Tonnage.
	inches.	inches.	inches.	ton.
Internal	19·062	7·062	5·625	..
External (over all) ..	21·812	8·312	6·937	·01819

11317—Cartridge, S.A., blank, .303-inch, cordite,   24 Dec 1901
    with mock bullet. (Mark VI)   C   6 Jan 1902
    Solid case.   19 Aug 1902

A pattern (design R.L. 11263) of the above-mentioned cartridge has been sealed to govern future manufacture.

The cartridge (see accompanying drawings) differs from the Mark V (LoC 7519) in having a mock bullet, which admits of the cartridge being loaded through the magazine and carried in bandoliers.

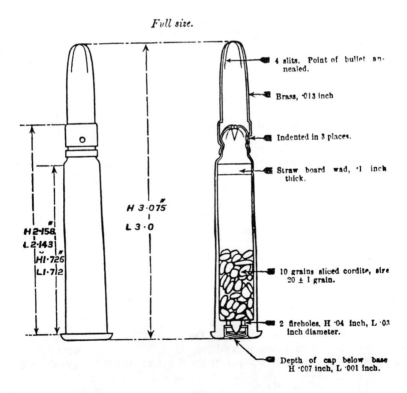

Full size.

The mock bullet is composed of brass .013 inch thick, and is secured over the neck of the case by indents. The nose is annealed, and has four slits. The case is blackened for further identification.

The existing stock of earlier pattern blank ammunition in land service will be used up by units equipped with pouches.

No. of cartridges	Packing Package	Weight (approx)
	(Naval and land service)	
1,400	Barrel, cartridge, quarter	54 lb.
2,400	Barrel, cartridge, half	90 lb.
	(Land service only)	
2,000	Box, A.S.A., 1,000 rounds, Mark XIII	79 lb.
1,100	Box, A.S.A., home and special, Mark XIV	46 lb.

11330—Sword, drummers. (Mark II)　　L　　13 Nov 1901
　　And band and buglers, Line regiments; brass hilt with buff piece.

Sealing of new pattern with "E.R." cypher.

A pattern of the above-mentioned sword has been sealed to govern future manufacture.

It differs from that described in LoC 7953 only in the substitution of "E.R." for "V.R." in the cypher on the crosspiece of the hilt. The numeral has not been advanced.

11332—Locket, union, gilt, Cadets. (Mark II)　　L　　2 Oct. 1902
　　Sandhurst.

A pattern of the above-mentioned locket has been sealed to govern future manufacture.

It differs from the previous pattern, referred to at p. 112, "Priced Vocabulary of Stores, 1902," but not published in List of Changes, in the Monogram and the design of the Crown.

11337—Frog, sword, saddle, R.A. (Mark I)　　8 Mar 1902
　　Leather, with strap for picketing peg.

Obsolete.

As swords are no longer to be carried by Royal Horse and Royal Field Artillery (A.O. 242, December 1901), except by warrant officers, the above-mentioned frog (LoC 9590) will cease to form part of the saddlery equipment for the same, and is hereby declared

obsolete.

All "Frogs, sword, saddle, R.A.," in possession of Royal Horse Artillery and Royal Field Artillery, except those in possession of warrant officers, will be converted to "Frogs, sword, saddle, Cavalry, Line (Mark III)" (LoC 6671), by removing the leather piece, strap, and buckle added by LoC 9590.

11397—Cartridges, S.A., ball, .303-inch, cordite—   18 Nov 1902
    Marks IV and V   C
1. Obsolete.

    Mark II   C
2. Continued for future manufacture.

1. The Marks IV and V .303-inch S.A. ball cordite cartridges (LoC 9089 and 9861) are hereby declared obsolete, the existing stock to be used up for practice.

2. The Mark II cartridge (LoC 7278) will continue to be manufactured in future.

11398—Cartridges, S.A. ball, pistol,   18 Nov 1902
    Webley, cordite—   26 Nov 1902
    Mark III   C
    Also Enfield.
1. Obsolete.

    Mark II   C
2. Continued for future manufacture.

1. With reference to LoC 10273, the Mark III cartridge is hereby declared obsolete, and existing stock will be used up as directed in LoC 10273.

2. The Mark II cartridge (LoC 9159, 10273) will continue to be manufactured in future.

11413—Cartridge, S.A. blank, .303-inch,   8 Sep 1902
    black powder, without bullet. (Mark III) N

No more to be made.

Consequent upon the adoption for naval service of the "Cartridge, S.A., blank, .303-inch, cordite, with mock bullet, Mark VI," (LoC 11317), no more cartridges of the above-mentioned description (LoC 9352) will be made, and so soon as the existing stock is used up, the pattern will be regarded as obsolete.

11489—Gloves, fencing, canvas, pair. (Mark II)   3 Oct 1902
    Bayonet.
    Musket, fencing. (Mark VI)
    Wood, with spring bayonet and india-rubber ball.
    Pad, fencing, body, canvas. (Mark II)
    With waist strap; bayonet.

    Patterns of the above-mentioned articles have been sealed to govern future manufacture. Existing gloves, muskets and pads will be used up.

    They differ from the previous patterns, LoC 6664 and 9982, in the following particulars—

    The gloves are not fitted with holes in the palms, they are also lighter in weight, being 1 lb. 5 oz. per pair instead of 1 lb. 8½ oz.

    The musket consists of a butt portion or stock of ash, 3 feet 9 inches in length, and bored at the barrel end to a depth of 2 feet 7½ inches into which is fitted a steel tube, threaded on the inside for the purpose of securing a brass cap, which in turn holds a spiral spring, 2 feet 6 inches long (when released), in position and admits of the free movement of the bayonet portion in the steel lining.

    The bayonet is made of hickory wood and is 19 inches long and 7/8 inch in diameter, and is kept in position in the stock by a piece of steel tubing secured to the back end, which butts against the steel lining of the brass cap when not compressed in use.

    On the front end of the bayonet a steel cap, 1¾ inches in diameter is secured which holds the indiarubber ball, which is also 1¾ inches in diameter, the latter is covered with linen and secured to the end of the bayonet with string.

    The pads are hollowed out at the sides so as to fit close into the thighs, they are also lighter in weight, viz., 1 lb. 1½ oz. each instead of 1 lb. 6½ oz.

11498—Carbines—   6 Sep 1902
    Artillery, Martini-Metford.   3 Feb 1903
      (Marks II and III)   L
    Cavalry, Martini-Metford.
      (Marks I*, II* and III)   L

    Rifles—
    Magazine Lee-Metford (Marks II and II*) C

Fitting with barrels having Enfield rifling.
Nomenclature when so fitted.

With reference to LoC 9124, 10439 and 11078— Arms of the above-mentioned description will, when requiring new barrels, be fitted with barrels having Enfield rifling. Such barrels will have the letter "E" marked on the knox-form.

The arms that are similar in exterior form and sighting to the arms with Metford barrels, viz.—
    Carbines, Cavalry, Martini-Metford (Marks I* and II*).
    Rifles, magazine Lee-Metford (Mark II),
will retain their present nomenclature when fitted with the Enfield barrel.

The remaining arms which are different will, on the fitting of the Enfield barrel, have their nomenclature altered as follows—
from Carbine, Artillery, Martini-Metford (Mark II) L
  to Carbine, Artillery, Martini-Enfield (Mark III) L
from Carbine, Artillery, Martini-Metford (Mark III) L
  to Carbine, Artillery, Martini-Enfield (Mark I) L
from Carbine, Cavalry, Martini-Metford (Mark III) L
  to Carbine, Cavalry, Martini-Enfield (Mark I) L
from Rifle, magazine, Lee-Metford (Mark II*) C
  to Rifle, magazine, Lee-Enfield (Mark I) C
(when fitted with old pattern fore-end and nose-cap, which are grooved for cleaning rod) or—
  to Rifle, magazine, Lee-Enfield (Mark I*) C
(when fitted with new pattern solid fore-end and nose-cap).

The old marking on the bodies of the arms fitted with new barrels will be cancelled and the new nomenclature stamped on as above. At present the marking will be done in Ordnance Factories, but in future if barrels are fitted locally the necessary stamps for marking the arms should be obtained as follows—
For land service arms: from the Chief Ordnance Officer, Weedon.
For naval service arms: supply to Naval Ordnance Depots is being arranged by the Superintendent of Ordnance Stores.

11632—Boxes, ammunition, S.A.—		7 Jan 1903
Pistol, Enfield, 240 rounds	L	18 Feb 1903
.303-inch, half, naval	N	
G. S.	C	
Home and special	L	
Mark XV	L	
750 rounds, .303-inch	L	

Modification to sliding lid.
New pattern of closing plate.

In future manufacture, small-arm ammunition boxes of the above-mentioned descriptions will be modified as follows—

1. The sliding lid will be tapered on both edges, instead of one as hitherto, and the tongues will be reduced in length from 1¾ inches to 1 inch to prevent undue "sticking".

2. The closing plate will be of slightly different shape so as to facilitate its removal when opening the tin lining.

11659—**Accoutrements, naval, pattern 1901**— 13 Jan 1903
    Bandolier (Mark II)   N   11 Feb 1903
    Brown, leather, 60 rounds, with   4 Apr 1903
    steadying strap.   13 May 1903
    Carriage, waterbottle. (Mark II)   N
    Brown leather, with shoulder strap.
    Cover, mess-tin. (Mark I)   N
    Brown canvas, with mess-tin strap
    sewn on, and 4 leather loops.
    Frog. (Mark I)   N
    Brown leather, with shifting loop;
    sword-bayonet or cutlass.

1. New patterns.

    Strap, mess-tin. (Mark I)   N

2. Obsolete as a separate article.

1. Patterns of the above-mentioned articles have been sealed to govern future manufacture. They differ from the previous patterns (LoC 11110) in the following particulars—

The Bandolier is cut on the curve to enable it to fit better on the wearer's shoulder, and the steadying strap has been riveted to the dee.

The Carriage, water-bottle, has the shoulder strap slightly altered to prevent it from being detached from the carriage.

The Cover, mess-tin, has the mess-tin strap sewn on it; existing stocks are to be altered locally, a sample to guide alteration being issued to naval depots for that purpose. The numeral has not been advanced in this case.

The Frog has the lower portion reduced in width 5/8 inch to fit the scabbard more tightly.

2. Consequent on the addition of the mess-tin strap to the cover, this strap becomes obsolete as a separate article, and instructions as to disposal will be issued by the Superintendent of Ordnance Stores.

11660—Cover, cuirass. (Mark I) L   23 Mar 1903
                                     14 May 1903

Fitting with undyed canvas division.

In future manufacture, the above-mentioned cover (LoC 7271) will be fitted with an undyed canvas division to preserve the cuirass from rust. The sealed pattern has been altered accordingly.

Covers in Ordnance charge will be altered from time to time as required, under arrangements which will be made by the Principal Ordnance Officer; those in possession of units will be dealt with in accordance with instructions which will be issued to Officers Commanding by the Principal Ordnance Officer.

11713—Carbines—		19 Sep 1902
Martini-Enfield—		20 Nov 1902
Artillery	L	16 Jan 1903
Cavalry	L	27 Mar 1903
Martini-Metford—		21 Apr 1903
Artillery	L	
Cavalry	L	

Removal of protecting wings on foresight.

Approval has been given to omit the protecting wings on the foresights of carbines, Martini-Enfield, artillery and cavalry, in future manufacture, and also to omit the wings for barrels for the repair of carbines, Martini-Metford, artillery and cavalry.

Carbines of the above patterns in the hands of troops at Sierra Leone will have the protecting wings removed locally, in accordance with the following instructions—
   Grind off the sharp corners of the safe edge of the file used, to prevent damaging the barleycorn, commence by filing the outer sides of the wings down to the neck of the block, then file off the top of the wings down to the base of the barleycorn, round off front corners, and smooth up.

Protectors, front sight (LoC 11714), No. 1 for carbines, artillery, M.M. and M.E., and No. 2 for carbines, cavalry, M.M., Mark III, and M.E. will be used with the above patterns of carbines when the wings have been removed.

11714—Protectors, front sight—		22 Dec 1902
No. 1	C	4 Feb 1903
Steel; rifles, magazine; and carbines,		20 Nov 1902
M.M. and M.E.		16 Jan 1903
No. 2	L	26 Feb 1903
Steel; carbines, cavalry, M.M., Mark III, and M.E.		14 Mar 1903

Patterns of the above-mentioned sight protectors, as shown in the accompanying drawings, have been approved to govern future manufacture.

The No. 1 protector for magazine rifles, and carbines, M.M., and M.E. artillery, differs from the protector front sight, "pattern C," described in LoC 9358, 9537, in being strengthened by the lap in front of hood being extended to bear on the barrel, and in having the lap and join brazed. The gap through which the front sight block passes, when the protector is put on, is reduced, to prevent the protector being removed, except by the use of the tool, armourers, removing and replacing protectors, front sight, Nos. 1 and 2 (LoC 11718).

The No. 2 protector for carbines, cavalry, M.M., Mark III, and M.E., is similar to the No. 1, but is of smaller size. It has been designed for the protection of the front sights of these carbines when the protecting wings have been removed (LoC 11713), and also to prevent the protectors being removed without the use of the tool, armourers, removing and replacing, &c. (LoC 11718).

Full size.
Protectors, front sight.
No. 1.
To be marked here.

No. 2.

To be marked here.

The protectors will be marked on the bottom, under the hood, with consecutive numbers by each unit.

11715–Rifle, short, magazine, Lee-Enfield.   23 Dec 1902
       (Mark I)   C   3 Jul 1903
                                                                               1 Jul 1903

A pattern of the above-mentioned rifle, as shown in the accompanying drawing, has been sealed to govern future manufacture.

The rifle is about 1¼ lb. lighter, and 5 inches shorter than the "Rifle, M.L.E., Mark I," (LoC 8117).

The magazine will hold 10 rounds, and is filled by cartridges carried in chargers (LoC 11753); guides to hold the latter are provided on the bolt head and body, while the five cartridges held in the charger are swept out of it by the thumb into the magazine. The bolt and cocking piece may be locked in the "full cock" and "fired" positions by a safety catch, and locking bolt, situated on the left side of the body, between it and the aperture sight; they are held in position by the aperture sight spring.

Barrel– The barrel is smaller in diameter externally, and 5 inches shorter than that of the L.E. rifle; it is fitted with a band which carries the block foresight, this is keyed and pinned to the barrel, and is dovetailed to carry the adjustable barleycorn foresight (three heights of which will be provided, known as high, normal and low, and marked "H", "N", and "L", respectively), to enable variations in shooting to be corrected before the rifle is issued to the Service. The heights of the high and low barleycorns differ from the normal by .015 inch.

Backsight– The backsight is fitted with a leaf pivoted to the bed at the front end; at the rear end is a cap in which the V is cut. Elevation is given by moving the slide, which rests on curved ramps on each side of the bed. The leaf is graduated by lines for every 100 yards from 200 to 2,000, the even numbers being marked by figures. The slide can be set at any 100 yards graduation or intermediate 50 yards elevation, and is held in position by means of catches engaging in notches on each side of the leaf. To set the slide, these catches are disengaged from the notches by pressing the bone studs at each side of the slide. The cap is joined to the leaf by a vertical dovetail, and it can be given a fine adjustment for intermediate ranges between the 50 yards elevation afforded by the slide by means of a vertical screw underneath the cap of the backsight; a small vernier scale divided to give a vertical movement of .0106 inch being provided on the left edge of the cap and leaf. Each division on the vernier represents 2 inches elevation per 100 yards. By raising the slide to its highest limit, 2,050 yards elevation can be obtained. The dial sight for long ranges is graduated from 1,600 to 2,800 yards.

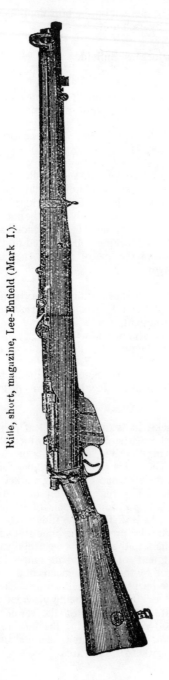

Rifle, short, magazine, Lee-Enfield (Mark I.).

The following particulars show the principal differences in detail between the short rifle and the M.L.M. and L.E. rifles—

Body— The body is made with charger guide on the left to receive the charger by which the magazine is loaded, and a stop on the right which forces the charger guide on the bolt head forward when the bolt is drawn fully back; it is also arranged to receive the locking bolt and safety catch to lock the bolt and cocking piece; the cut-off slot is left in the body for the insertion of a cut-off if required. The left side of the body is cut away to afford clearance for the thumb of the right hand when pressing the cartridges from the charger into the magazine. A cut-off will be supplied for naval service only.

Bolt— The bolt rib is lower, and the handle is set closer to the body; the bolt cover and extension for the safety catch is omitted.

Bolt head— The bolt head is made with a slide for the bolt head charger guide, and has a slot cut in the screwed end of the bolt head, which acts as a key when stripping and assembling the striker and cocking piece.

Cocking piece— The cocking piece is shorter, and is locked by a locking bolt, the point of which, when the thumb piece is turned back, enters recesses on the left side of the cocking piece and locks it in the full cocked and fired positions. The screw keeper striker is replaced by a nut keeper striker, this is screwed on to a screw, round the shank of which is a spiral spring, contained in a recess in the cocking piece. The nut keeper striker may be pulled to the rear and slightly turned by the finger and thumb, the striker can then be unscrewed from the cocking piece by unscrewing the bolt head; the bolt is thus completely stripped without the aid of tools.

Trigger— The trigger is provided with two ribs, which bear in succession on the lower arm of the sear, and produce a double pull off. The strength of the first pull is 3 to 4 lb., and of the second, 5 to 6 lb.

Butt plate— The sheet steel butt plate is lighter, the butt trap, pin, spring, spring screw, strap and strap screw being omitted.

Magazine— The magazine is slightly deeper in the rear to give more room for the 10 cartridges, and so facilitate loading the second five cartridges from a charger. It is provided with a zigzag platform spring and auxiliary spring. It has a stop clip to keep the right hand cartridge in position, and to enable the platform and spring to be easily removed for cleaning.

Band, inner— The inner band, which encircles the barrel at the centre with .002-inch freedom, is fitted inside the fore-end, and is held in position by a screw, spiral spring, and washer, so that it supports the barrel, without holding it rigidly, or preventing expansion.

Band, outer— The outer band encircles the fore-end and handguard and inner band; it is jointed at the top, and held together by a screw underneath, which also carries the sling swivel. The swivel screws for the butt, band and nosecap, are interchangeable.

Nosecap— The nosecap is considerably larger than that of the L.E. rifle; its front end is flush with the muzzle of the barrel, and has an extension in front on which the crosspiece of the sword-bayonet fits, and a bar underneath the rear end to hold the pommel of the sword-bayonet; it is also provided with lugs to carry the swivel and screw, and has high wings to protect the foresight; the muzzle of the barrel has .002 inch freedom in the barrel hole.

Swivels— There are two swivels, one attached to outer band and one to butt, the latter swivel can be attached to lug on nosecap for use in slinging rifle on back when mounted. For Naval Service only, a piling swivel will be attached in this position.

Handguard— The handguard completely covers the barrel, extending from the body to the nosecap; it is in two pieces, being divided diagonally at the centre of the sight bed; the front portion is held in position by the outer band, and its front end fits in a recess in the nosecap; the rear portion is held in position by a spring gripping the barrel; both portions rest upon the shoulders of the fore-end, being quite free of the barrel throughout. In stripping, the rear handguard must be first removed, the front handguard can be pushed back clear of the nosecap after the outer band has been removed. The rear portion is fitted with a backsight protector, consisting of two upstanding ears, which protect the cap of the backsight when the latter is adjusted for short ranges.

Stock, fore-end— The stock, fore-end, extends to within 1/8-inch of the muzzle of the barrel; it is free in the barrel groove throughout, excepting about ½ inch in front and rear of the inner band, and under the knoxform at the breech end. It is fitted with a keeper plate let into the breech end, into which the squared end of the stock bolt fits, to prevent the stock bolt turning and the stock butt becoming loose.

Stock, butt— The stock, butt, is issued in three lengths, one ½ inch shorter and one ½ inch longer than the normal, and marked respectively S and L. It is fitted with a sheet steel butt plate without butt trap, as the oil bottle and pull-through are not carried in the butt. It is bored longitudinally with four holes for lightness, and is provided with a marking disc screwed into the right side. The stock bolt is shorter, and is squared at the front end to fit the keeper plate.

In stripping the rifle it is necessary that the fore-end should be first removed before turning the stock bolt.

Particulars relating to rifling, sighting, weight, &c.
Length of barrel . . . . .25 3/16 inches.
Calibre . . . . .303 inch.
Rifling. . . . . .Enfield
Grooves, number. 5
 depth at muzzle. . . . .0065 inch.
 depth at breech, to within
  14 inches of muzzle . . . .005 inch.
Width of lands . . . .0936 inch.
Twist of rifling, left-handed. . . . . .1 turn in 10 ins.
Sighting system. . . . . .Adjustable barley-
 corn front sight,
 radial back sight.
Distance between barleycorn
 and back sight, V. . . . . .1 ft. 7 5/32 ins.
Length of rifle . . . . .3 ft. 8 9/16 ins.
Length of rifle with sword-bayonet. . . . . . .4 ft. 8 11/16 ins.
Length of sword-bayonet (over all). . . . . . .1 ft. 4 7/8 ins.
Length of sword-bayonet blade . . . . .1 ft. 0 1/8 ins.
Weight of rifle, with magazine empty . . . . .8 lb. 2½ oz.
Weight of sword-bayonet . . . . .1 lb. 0½ oz.
Weight of sword-bayonet scabbard . . . . . . .0 lb. 4½ oz.

Ammunition same as for M.L.E., M.L.M., M.M., and M.E. arms.

The following components are special to this rifle—

Components

Barleycorn foresight—
 Normal
 High
 Low
 Barrel
Block, band, foresight
Body
Bolt
Bolt head
Bolt head charger guide
Butt plate
Band, outer
Band, inner
Bed, sight, back
Bolt, stock
Clip, stop, magazine
Cocking piece
Cut-off †
Extractor

Guard, trigger
Guard, hand, front
Guard, hand, rear
Guard, hand, front cap
Guard, hand, rear sight
 protector
Head, with catch, slide,
 sight leaf (2)
Key, block, band, foresight
Leaf, sight, back
Locking bolt
Locking bolt stop pins (2)
Locking bolt safety catch
Magazine case
Nosecap
Nut, keeper, striker (with pin)
Nut, screw, back, nosecap

† In rifles for naval service only.

Pin, axis, leaf, sight, back
Pin, fixing, bed, sight, back
Pin, fixing, block, band,
  foresight
Pin, fixing, stud, head, catch,
  slide, sight, back
Pin, joint, band, outer
Plate, dial, sight
Platform, magazine
Rivets, handguard, rear, sight
  protector, side
Rivets, handguard, cap and
  sight protector, top (3)
Spring, aperture sight
Springs, catch, slide,
  sight leaf (2)
Spring, nut, keeper, striker
Spring, platform, magazine
Spring, auxiliary, platform,
  magazine
Spring, screw, band, inner
Spring, sight, back
Screw, band, outer, and
  swivels, butt and piling
Screw, band inner
Screw, guard, back
Screws, bed, sight, back—
  Front
  Back
Screw, dial sight, fixing

Screw, fine adjustment, sight
  leaf
Screw, keeper, slide, fine
  adjustment, leaf, sight, back
Screws, nosecap—
  Back
  Front
Screw, nut, keeper, striker
Screw, spring, aperture sight
Screw, spring, sight, back
Screw, stop, charger guide
Sear
Slide, sight, back
Slide, fine adjustment, leaf,
  sight, back
Striker
Stem, swivel, butt
Stock, butt—
  Normal
  Long
  Short
Stock, fore-end
Stud, clip, stop, magazine
Trigger
Washer, spring, band, inner
Washer, pin, axis, sight, back
Washers, rivet, fore-end; cap,
  spring, and sight protector,
  handguard (7)

The pattern 1903 sword-bayonet (LoC 11716, 11717), is used with this rifle.

11716—Sword-bayonet, pattern 1903   C   19 Dec 1902
    For rifle, short, M.L.E.   3 Jul 1903

A pattern of the above-mentioned sword-bayonet has been sealed to govern future manufacture.

The sword-bayonet (see accompanying drawing) differs from the pattern 1888 (Mark III), LoC 11151, in the size and position of the mortising for the bar on nosecap of rifle, and in the figure of the pommel as shown in the accompanying drawing.

The bayonet is attached to the bar and stud on the nosecap of the "Rifle, short, M.L.E." and not on the bar on nosecap and muzzle of the barrel, as in Rifles M.L.M. and M.L.E.

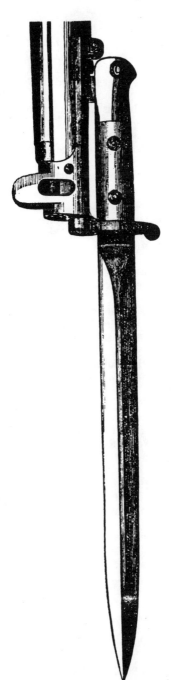
Sword-bayonet, pattern 1903.

Weight of sword-bayonet . . . . .1 lb. ½ oz.
Weight of sword-bayonet scabbard . . . . . . .4½ oz.
Length of sword-bayonet blade . . . . .12 1/8 ins.
Length of sword-bayonet over all . . . . .1 foot 4 7/8 ins.
Length of sword-bayonet scabbard . . . . . . .13¼ ins.

The following scabbards are interchangeable on the sword-bayonets, pattern 1903—
Scabbard, sword bayonet, pattern 1888 (Mark I) C
Scabbard, brown leather, sword-bayonet, pattern 1888, land, (Mark II) L
Scabbard, brown leather, sword-bayonet, pattern 1888, naval, (Mark I) N

**11717—Sword-bayonet, pattern 1903, converted   C   3 Feb 1903
from sword-bayonet, pattern 1888; for   26 Feb 1903
rifle, short, M.L.E.   3 Jul 1903**

A pattern of the above-mentioned sword-bayonet has been provisionally sealed to govern conversions as may be ordered.

The conversion is made from sword-bayonets, pattern 1888, Mark I, II and III, and consists generally in fitting a new pommel, arranged to fit the "Rifle, short, M.L.E."

After conversion the whole of the components are interchangeable with those of the new sword-bayonet, pattern 1903, LoC 11716.

**11718—Tools, armourers—   21 Apr 1903
Tool, removing and replacing protectors,
front sight, Nos. 1 and 2   C**
Steel, with handle.

A pattern of the above-mentioned tool for removing and replacing the protectors, front sight, Nos. 1 and 2 (LoC 11714) has been approved to govern manufacture.

Instructions for using the tool.
To take off protector— Draw back the protector until the hood is free of the front sight block, turn it until the gap at the spring end is in line with the block, then, holding the tool firmly by the handle, place the horns in the opening at the spring end of the protector, press firmly on the barrel, turn the tool to the right, so as to spring open the ends of the protector, and then draw off the protector.

To put on protector— Place the spring end of the protector on the barrel, gap in line with the front sight block, open as before, slide as far as it will go, turn and push over the block.

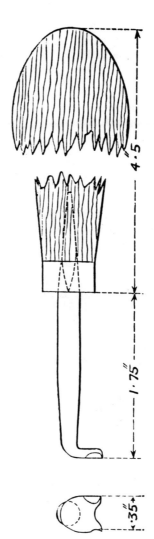

Tool, removing and replacing protectors, front sight, Nos. 1 and 2.
Full size.

11752—Cartridges, S.A. ball, M.H.—   18 Dec 1902
    Rifle, solid case, cordite. (Mark I)   L   25 Mar 1903
    Not to be used in machine guns.   31 Mar 1903

    Carbine, solid case, cordite. (Mark I)   L

Drawings (R.L. 8181c and 12553a respectively) of the above-mentioned cartridges have been sealed to govern manufacture for land service as may be authorized.

Rifle cartridge.

The case is the same as that used with the Mark II powder cartridge described in LoC 4911. The charge consists of 35.8 grains of cordite, size 3, on top of which is placed a waxed millboard wad.

The bullet weighs 480 grains, and is provided with two cannelures, the neck of the case being choked into the rear cannelure. The bullet is covered for about two-thirds up the body with paper which is coloured blue, to facilitate identification.

Certain issues have been made in which the bullets were provided with orange coloured paper, but no more will be issued so covered. The wrappers of these cartridges were stamped "For rifle only."

Carbine cartridge.

The case is the same as that used with the Mark I powder cartridge described in LoC 5159. The charge consists of 34 grains of cordite, size 3, on top of which is placed a waxed millboard wad. The bullet weighs 410 grains, and is provided with one cannelure, into which the neck of the case is choked. The bullet is covered for about two-thirds up the body with paper, which is coloured green to facilitate identification.

Packing.

Rifle cartridge; 580 rounds in Box, ammunition, S.A., G.S.
Carbine cartridge: 600 rounds in Box, ammunition, S.A., G.S.

11753—Charger, .303-inch cartridges. (Mark I)   C   16 Jan 1903
    Steel, to hold five.   25 Feb 1903

Drawing (C.I.W. 687) of the above-mentioned charger has been sealed to govern manufacture.

The charger, which is for use with the "Rifle, short, M.L.E.," is made of steel, to the form and dimensions shown on the accompanying drawings, and holds five .303-inch cartridges.

Charger, ·303-inch cartridges. (Mark I.)

Full size.

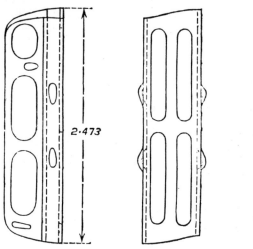

11807—Carbine, magazine, Lee-Enfield   L   7 Oct 1902
Fitted to take pattern 1888 sword-bayonet.

Sighting.

With reference to LoC 10220— The sighting of the above-mentioned carbine has been corrected by the substitution of a backsight leaf, with the cap lower and the slide narrower.

This alteration makes the leaf and slide special to this carbine. Spare leaves when issued will be marked "E.C. 88" on the right bottom corner.

11808—Chest, rifle, short, M.L.E. (Mark I)   C   18 Sep 1903
Wood, with fittings; to contain 20 rifles, sword-bayonets, and scabbards.

A pattern of the above-mentioned chest has been approved to govern future manufacture and conversion of existing pattern when ordered.

It is made of deal, with elm ends, and is provided with cleats and rope handles. The lid is secured with screws. The inside fittings are of deal, coated with paraffin wax.

Dimensions, &c.

Length, overall .50 3/8 inches.
Width, overall . . . . .21 1/8 inches.
Depth, overall . . . . .17 3/8 inches.
Tonnage . . . .2675 tons.

11810–Rifles, M.E.–   13 Feb 1903
    (Mark I*)   L
    (Mark II*)   L
    Both marks with adjustable foresights.

    Barleycorns, foresight, M.E. rifle–
      High. (Mark I)   L
      Normal. (Mark I)   L
      Low. (Mark I)   L
    for Marks I* and II* M.E. rifles.

With reference to LoC 8118 and 8196– A pattern rifle, M.E. (Mark II*) and three adjustable barleycorn foresights, have been sealed to govern repair and alteration when the latter is specially ordered.

Any rifles, M.E. (Marks I and II) which are selected for the adjustment of the sighting, will have suitable adjustable barleycorn foresights fitted at the Royal Small Arms Factory, Birmingham.

Three heights of foresight will be provided, known as high, normal and low, and marked "H", "N" and "L", respectively. The heights of the high and low differ from the normal by .015 inch.

Rifles, M.E., when fitted with adjustable barleycorn foresights, will be marked with a star (*) on the right side of the numeral on the body.

Barleycorn, foresight, M.E. rifles.
Full size.

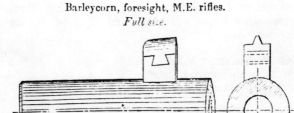

11811—Scabbard, sword-bayonet, pattern 1903,   25 Aug 1903
land. (Mark I)   L   27 Aug 1903
Brown leather; also sword-bayonets,
pattern 1888.

A pattern of the above-mentioned scabbard, as shown in the accompanying drawing, has been sealed to govern future manufacture. *(See page 66 for drawing).*

It differs from the "Scabbard, brown leather, sword-bayonet, pattern 1888, land (Mark II)", LoC 11151, in the following particulars—
    A steel chape is fitted inside the bottom end of the leather.
    The leather is blocked flesh side out instead of grain side out, to facilitate repair.
    The tag is made in the form of a loop, so that it may be hung on the waistbelt.

The scabbard is interchangeable on sword-bayonets, pattern 1888 (Marks I, II and III); pattern 1903 (Mark I) and converted.

    Length of scabbard overall . . . . .18 1/8 to 18 3/8 in
    Weight of scabbard . . . . .7 to 8 oz.

11812—Scabbard, sword-bayonet, pattern 1903,   3 Jun 1903
naval. (Mark I)   N
Brown leather; also sword-bayonets,
pattern 1888.

A pattern of the above-mentioned scabbard, as shown in the accompanying drawing has been sealed to govern future manufacture.

Scale ¼.

13" MAX.
12 3/4" MIN.

It differs from the "Scabbard, brown leather, sword-bayonet, pattern 1888, naval (Mark I)" (LoC 11109), in the following particulars—
    A steel chape is fitted inside the end of the leather.
    The leather is blocked flesh side out, instead of grain side out, to facilitate repair.

The scabbard is interchangeable on sword-bayonets, pattern 1888 (Marks I, II and III); pattern 1903 (Mark I) and (converted).
    Weight. . . . . .4½ to 5½ oz.

Scabbard, sword-bayonet, pattern 1903, land. (Mark I.)
Scale ⅓.

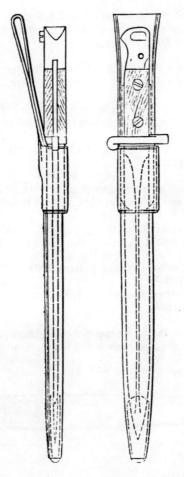

LoC 11811

11832—Cartridge, S.A., dummy, drill, .303-inch,   29 May 1903
  rifles or carbines (Mark III)   L

A pattern (design R.L. 12859) of the above-mentioned cartridge has been approved to govern future manufacture for land service.

It differs from the previous pattern (LoC 9519) in the bullet being made of boxwood. The bullet is fixed in the case by three indents and by the mouth of the case being coned.

11849—**Case, 200 scabbards, sword-bayonet,**   16 Oct 1903
pattern 1903, land. (Mark I).   L
Wood, with fittings; also 220 scabbards, sword-bayonet, pattern 1888.

A pattern of the above-mentioned case has been approved to govern manufacture for land service.

It is made of deal, with elm ends, and is provided with fittings to carry either 200 "Scabbards, sword-bayonet, pattern 1903", (LoC 11811) or 220 "Scabbards, sword-bayonet, pattern 1888", as may be required.

The inside fittings are of deal, coated with paraffin wax.

A cleat, with rope handle for lifting, is attached to each end of the case. The lid is secured by screws.

Dimensions, &c., (for stowage).
Length, overall . . . . .4 ft. 0 3/8 ins.
Width, overall. . . . . .1 ft. 6 1/16 ins.
Depth, overall. . . . . .1 ft. 2 3/16 ins.
Tonnage . . . .17934 ton.

11850—**Cut-off, rifle, short, M.L.E. (Mark I)**   N   12 Aug 1903
and rifle, short, M.L.E., converted, Mark II   18 Aug 1903

A pattern of the above-mentioned cut-off (as shown in the accompanying drawing) has been sealed to govern manufacture for "Rifles, short, M.L.E. Mark I and converted, Mark II" for naval service only.

Cut-off, rifle, short, M.L.E.
Full size.

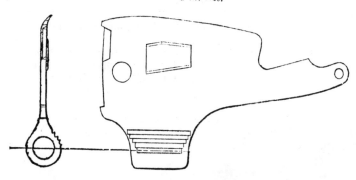

It differs from the "Cut-off, rifle, M.L.E., Mark II" (LoC 7404) in the following particulars—
A space has been cut near the joint to clear the magazine stop clip, and the left edge has been turned up to facilitate cutting off the cartridges in the magazine.

11851—Tools and materials, armourers—   10 Aug 1903
Rod, cleaning, .303-inch, No. 4 (Mark I)   C
With swivel handle; for use with brass wire; rifle, short, M.L.E.
Bush, stop, rod, cleaning, .303-inch, No. 2 (Mark I)   C
With screw; rifle, short, M.L.E.

1. New patterns.

Muzzle guide, carbine, cavalry, .303-inch and rifle, short, M.L.E.   C
For use with cleaning rod.

2. Nomenclature.

1. Patterns of the above-mentioned tools (a drawing of which is attached) have been sealed to govern manufacture for use with rifles, short, M.L.E.

The "Rod, cleaning, No. 4" is similar to the rods, No. 2 and 3 (LoC 11079), but is 5 inches shorter to suit the barrels of the short rifles.

The "Bush stop" is 5 inches long, and is made to fit on the "Rod, cleaning, No. 2" (LoC 11079), to enable the rod to be used for the barrels of the short rifles.

The No. 2 rod will be replaced by the No. 4 as the stock of the former is used up.

When troops are armed with short rifles, officers concerned will demand from Army Ordnance Department, either "Bush, stop, rod, cleaning," or "Rod, cleaning, No. 4" as may be necessary.

Instructions for fitting the "bush, stop," to the No. 2 rod.
Remove the leather washer from the rod, place the bush on the rod (small end first) up to the collar of the rod, fix the bush by the screw, and replace the leather washer.

2. The use of the "Muzzle, guide, carbine, cavalry, .303-inch" (LoC 11079) has been extended to "Rifles, short, M.L.E." and the nomenclature has accordingly been amended as shown above.

Tools and materials, armourers.
Scale ½.

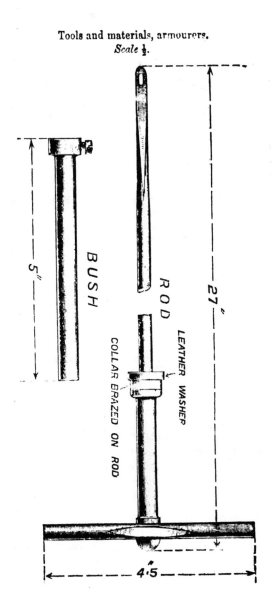

When using the muzzle guide with short rifles, the nosecap, band, fore-end, and handguards, &c., must first be removed.

11856—Bucket, rifle. (Mark IV) 31 Jul 1903
   Leather, with steadying and two suspending straps.

   Attachment to saddle.

   In the case of iron and steel arch saddles that are without supports (i.e. struts) to the hind arch, the above bucket (LoC 11208) will be suspended by passing one of the suspending straps through the loop of the other, forming one strap, which should then be passed round the hind arch of the saddle and buckled to the bucket.

11947—Rifle, short, magazine, Lee-Enfield 14 Sep 1903
   (Mark I) C 6 Nov 1903
   23 Dec 1902
   LoC 11715 cancelled. 3 Jul 1903
   1 Jul 1903

   Consequent on the addition of a wind-gauge and its components to the backsight, and the replacement of the rivets, handguard, cap, and sight protector by screws, LoC 11715 is hereby cancelled, and the following is published in substitution—

   A pattern of the above-mentioned rifle, as shown in the accompanying drawings, has been sealed to govern future manufacture.

   The rifle is about 1¼ lb. lighter, and 5 inches shorter than the "Rifle, M.L.E., Mark I", Loc 8117.

   The magazine will hold 10 rounds, and is filled by cartridges carried in chargers, LoC 11753; guides to hold the latter are provided on the bolt head and body, while the five cartridges held in the charger are swept out of it by the thumb into the magazine. The bolt and cocking piece may be locked into the "full cock" and "fired" positions by the safety catch, and locking bolt, situated on the left side of the body, between it and the aperture sight; they are held in position by the aperture sight spring.

   Barrel— The barrel is smaller in diameter externally, and 5 inches shorter than that of the L.E. rifle; it is fitted with a band which carries the block foresight, this is keyed and pinned to the barrel, and is dovetailed to carry the adjustable barleycorn foresight (three heights of which will be provided, known as high, normal, and low, and marked "H", "N" and "L" respectively), to enable variations in shooting to be corrected before the rifle is issued to the Service. The heights of the high and low barleycorns differ from the normal by .015 inch.

   Backsight— The backsight is fitted with a leaf pivoted to the bed at the front end; at the rear end a vertical cap and horizontal

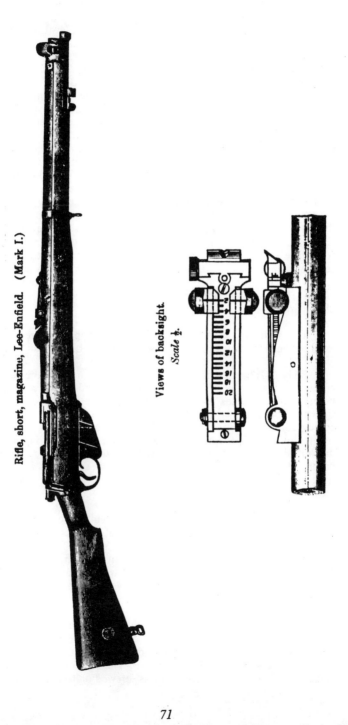

Rifle, short, magazine, Lee-Enfield. (Mark I.)

Views of backsight.
Scale ½.

wind-gauge are fitted, a V being cut in the latter. Elevation is given by moving the slide, which rests on curved ramps on each side of the bed. The leaf is graduated by lines for every 100 yards from 200 to 2,000, the even numbers being marked by figures. The slide can be set at any 100 yards graduation or intermediate 50 yards elevation, and is held in position by means of catches engaging in notches on each side of the leaf. To set the slide these catches are disengaged from the notches by pressing the bone studs at each side of the slide. The cap is joined to the leaf by a vertical dovetail, and it can be given a fine adjustment for intermediate ranges between the 50 yards elevation afforded by the slide by means of a vertical screw underneath the cap of the backsight; a small scale with divisions of .0318 inch being provided on the left edge of the cap and leaf. The wind-gauge fitted in the vertical cap is moved horizontally by means of a screw; a scale is marked on the wind-gauge and cap with divisions of .0318 inch. Each division on the wind-gauge and fine elevation adjustment represents 6 inches per 100 yards. A slot is cut in the top front edge of the cap for the thumb nail to be inserted to act as a stop to facilitate the centreing of the wind-gauge in the dusk. By raising the cap to its highest limit, 2,050 yards elevation can be obtained. The dial sight for long ranges is graduated from 1,600 to 2,800 yards.

The following particulars show the principal differences in detail between the short rifle and the M.L.M. and L.E. rifles—

Body— The body is made with charger guide on the left to receive the charger by which the magazine is loaded, and a stop on the right which forces the charger guide on the bolt head forward when the bolt is drawn fully back; it is also arranged to receive the locking bolt and safety catch to lock the bolt and cocking piece; the cut-off slot is left in the body for the insertion of a cut-off if required. The left side of the body is cut away to afford clearance for the thumb of the right hand when pressing the cartridges from the charger into the magazine. A cut-off will be supplied for naval service only.

Bolt— The bolt rib is lower, and the handle is set closer to the body; the bolt cover and extension for the safety catch is omitted.

Bolt-head— The bolt-head is made with a slide for the bolt-head charger guide, and has a slot cut in the screwed end of the bolt-head, which acts as a key when stripping and assembling the striker and cocking piece.

Cocking-piece— The cocking-piece is shorter, and is locked by a locking bolt, the point of which, when the thumb piece is turned back, enters recesses on the left side of the cocking piece and locks it in the full cocked and fired positions. The screw keeper striker is replaced by a nut keeper striker, this is screwed on to a screw, round the shank of which is a spiral spring, contained in a recess in the cocking piece. The nut keeper striker may be pulled to the rear and slightly turned by the finger and thumb, the striker can then be unscrewed from the cocking piece by unscrewing the bolt-head; the bolt is thus completely stripped without the aid of tools.

Trigger— The trigger is provided with two ribs, which bear in succession on the lower arm of the sear, and produce a double pull off. The strength of the first pull is 3 to 4 lb., and of the second, 5 to 6 lb.

Butt plate— The sheet steel butt plate is lighter, the butt trap, pin, spring, spring screw, strap and strap screw being omitted.

Magazine— The magazine is slightly deeper in the rear to give more room for the 10 cartridges, and so facilitate loading the second five cartridges from a charger. It is provided with a zigzag platform spring and auxiliary spring. It has a stop clip to keep the right hand cartridge in position, and to enable the platform and spring to be easily removed for cleaning.

Band, inner— The inner band, which encircles the barrel at the centre with .002-inch freedom, is fitted inside the fore-end, and is held in position by a screw, spiral spring, and washer, so that it suppports the barrel, without holding it rigidly, or preventing expansion.

Band, outer— The outer band encircles the fore-end and handguard and inner band; it is jointed at the top, and held together by a screw underneath, which also carries the sling swivel. The swivel screws for the butt, band, and nosecap, are interchangeable.

Nosecap— The nosecap is considerably larger than that of the L.E. rifle; its front end is flush with the muzzle of the barrel, and has an extension in front on which the crosspiece of the sword-bayonet fits, and a bar underneath the rear end to hold the pommel of the sword-bayonet; it is also provided with lugs to carry the swivel and screw, and has high wings to protect the foresight; the muzzle of the barrel has .002 inch freedom in the barrel hole.

Swivels— There are two swivels, one attached to outer band and one to butt, the latter swivel can be attached to lug on nosecap for use in slinging rifle on back when mounted. For Naval Service only, a piling swivel will be attached in this position.

Handguard— The handguard completely covers the barrel, extending from the body to the nosecap; it is in two pieces, being divided diagonally at the centre of the sight bed; the front portion is held in position by the outer band, and its front end fits in a recess in the nosecap; the rear portion is held in position by a spring gripping the barrel; both portions rest upon the shoulders of the fore-end, being quite free of the barrel throughout. In stripping, the rear handguard must be first removed, the front handguard can then be pushed back clear of the nosecap after the outer band has been removed. The rear portion is fitted with a backsight protector, consisting of two upstanding ears, which protect the cap of the backsight when the latter is adjusted for short ranges.

Stock, fore-end— The stock, fore-end extends to within 1/8 inch of the muzzle of the barrel; it is free in the barrel groove throughout, excepting about ½ inch in front and rear of the inner band, and under the knoxform at the breech end. It is fitted with a keeper plate let into the breech end, into which the squared end of the stock bolt fits, to prevent the stock bolt turning and the stock butt becoming loose.

Stock, butt— The stock, butt, is issued in three lengths, one ½ inch shorter and one ½ inch longer than the normal, and marked respectively S and L. It is fitted with a sheet steel butt plate without butt trap, as the oil bottle and pull-through are not carried in the butt. It is bored longitudinally with four holes for lightness, and is provided with a marking disc screwed into the right side. The stock bolt is shorter, and is squared at the front end to fit the keeper plate. In stripping the rifle it is necessary that the fore-end should be first removed before turning the stock bolt.

Particulars relating to rifling, sighting, weight, &c.,

Length of barrel . . . . .25 3/16 inches.
Calibre . . . .303 inch.
Rifling. . . . . .Enfield.
Grooves, number. 5
   depth, at muzzle . . . .0065 inch.
   depth, at breech, to within
   14 inches of the muzzle. . . . .005 inch.
Width of lands . . . .0936 inch.
Twist of rifling, left-handed. . . . . .1 turn in 10 ins.
Sighting system. . . . . .Adjustable barleycorn foresight, radial backsight, with vertical adjustment and wind-gauge.
Distance between barleycorn
   and backsight, V . . . . .1 ft. 7 13/32 ins.
Length of rifle . . . . .3 ft. 8 9/16 ins.
Length of rifle with sword-bayonet. . . . . . .4 ft. 8 11/16 ins.
Length of sword-bayonet, overall . . . . .1 ft. 4 7/8 ins.
Length of sword-bayonet blade . . . . .1 ft. 0 1/8 ins.
Weight of rifle, with magazine empty . . . . .8 lb. 2½ oz.
Weight of sword-bayonet . . . . .1 lb. 0½ oz.
Weight of sword-bayonet scabbard . . . . . . .0 lb. 4½ oz.

Ammunition same as for M.L.E., M.L.M., M.M. and M.E. arms.

The following components are special to this rifle—

Components.

Band, inner
Band, outer
Barleycorns foresight—
  High
  Low
  Normal
Barrel
Bed, sight, back
Block, band, foresight
Body
Bolt
Bolt head
Bolt head charger guide
Butt plate
Bolt, stock
Clip, stop, magazine
Cocking piece
Cut-off †
Extractor
Guard, hand, front
Guard, hand, front cap
Guard, hand, rear
Guard, hand, rear, sight protector
Guard, trigger
Head, with catch, slide, sight leaf (2)
Key, block, band, foresight
Leaf, sight, back
Locking bolt
Locking bolt safety catch
Locking bolt stop pins (2)
Magazine case
Nosecap
Nut, keeper, striker (with pin)
Nut, screw, back, nosecap
Pin, axis, sight, back
Pin, fixing, bed, sight, back
Pin, fixing, block, band, foresight
Pin, fixing, stud, head, catch, slide, sight, back (2)
Pin, fixing, washer, pin, axis, sight, back
Pin, joint, band, outer
Plate, dial, sight
Plate, keeper, screw, wind-gauge

Platform, magazine
Screw, dial sight, fixing
Screw, fine adjustment sight leaf
Screw, guard, back
Screw, nut, keeper, striker
Screw, protector, sight, side, handguard, rear
Screw, spring, sight, aperture
Screw, spring, sight, back
Screw, stop, charger guide
Screw, wind gauge
Screw, band, inner
Screws, band, outer, and swivels, butt and piling (3)
Screws, bed, sight, back—
  Front
  Back
Screws, keeper, fine adjustment, sight, back and handguard, cap and sight protector, top (4)
Screws, nosecap—
  Back
  Front
Sear
Slide, sight, back
Slide, sight, back, fine adjustment
Slide, sight, back, wind gauge
Spring, auxiliary, platform, magazine
Spring, nut, keeper, striker
Spring, platform, magazine
Spring, screw, band, inner
Spring, screw, wind gauge
Spring, sight, aperture
Spring, sight, back
Springs, catch, slide, sight leaf (2)
Stem, swivel, butt

† *In rifles for naval service only.*

Stocks, butt–
  Long
  Normal
  Short
Stock, fore-end
Striker

Stud, clip, stop, magazine
Trigger
Washer, pin, axis, sight, back
Washer, rivet, fore-end; and spring, handguard (4)
Washer, spring, band, inner

Bayonet.

The pattern 1903 sword bayonet, LoC 11716, 11717, is used with this rifle.

11948–Rifle, short, magazine, Lee-Enfield, converted (Mark I)   2 Nov 1903   L
Converted from Rifle, magazine, Lee-Metford Mark I*.

A pattern of the above-mentioned rifle has been sealed to govern conversion as may be ordered.

The conversion is made from rifles, M.L.M., Mark I*, and consists generally in fitting the rifles with new sights, shorter and lighter barrels, and adapting them for the magazine to be filled from a charger, LoC 11753.

The arm is fitted to carry the sword-bayonet, pattern 1903, LoC 11716, 11717, on the nose cap. The barrel and sights are identical with those described in LoC 11947.

Body– The body of the rifle is altered to receive the magazine, rifle, short, M.L.E., Mark I, the charger and the new locking and safety bolts.

Bolt– The bolt has the cover stops and a portion of the top of the rib removed, the cover screw holes filled in, and hole bored for screw, keeper, screw head, bolt breech.

Bolt head– A slot is cut in the tang of the bolt head, which acts as a key when stripping and assembling the striker and cocking piece; the screw, keeper, screw bolt head, and screw, bolt head, first being removed.

Dial sight– The dial sight plate has the original lining removed, and is relined and marked. The dial sight pointer is fitted with a new bead.

Trigger guard– The guard is expanded to receive the magazine, Rifle, short, M.L.E., Mark I. A piece is brazed in the magazine slot for magazine seating, the swivel boss is removed, and a new link loop is fitted to receive the link, Rifle M.L.M., Mark II; it is also lightened.

Stock butt— The stock butt is reduced externally, and bored longitudinally with three holes for lightness; it is arranged to take the butt plate and stock bolt, rifle, short, M.L.E., Mark I. When the stock butt is small at the socket end, the shoulders and end are cut back 3/16 inch, this shortens the converted rifle, as compared with the new, by about 1/4 inch. In packing converted rifles, an additional packing piece, 1/4 inch thick, will be required, inserted at the end of the chest behind the bridge fittings.

Stock, fore-end— The stock, fore-end, has a liner fitted and glued in; it is opened to receive the altered body; the clearing rod slot and the band pin hole are filled up. It is arranged to take the stock bolt keeper plate.

All dimensions, weights, &c., are the same as for rifle, short, M.L.E., Mark I, LoC 11947, and converted, Mark II, LoC 11949. The following components are special to this rifle—
Body †
Bolt †
Bolt head
Stock, butt †
Stock, fore-end †
Screw, bolt head
† Components marked thus are converted from rifle, M.L.M., Mark I*.

All other components are common to rifle, short, M.L.E., Mark I, LoC 11947, and rifle, short, M.L.E., converted, Mark II, LoC 11949.

11949—Rifle, short, magazine Lee-Enfield,   16 Jan 1903
    converted (Mark II)   C   14 Oct 1903
    Converted from rifles, magazine, L.E., Marks   6 Nov 1903
       I and I*, and magazine, L.M., Marks II and II*

A pattern of the above-mentioned rifle has been sealed to govern conversion as may be ordered. The rifle is very similar to the "Rifle, short, magazine, Lee-Enfield", LoC 11947. The conversion is made from rifles, M.L.E., Marks I and I*, and M.L.M., Marks II and II*, and consists generally in fitting the rifles with new sights, shorter and lighter barrels, and adapting them for the magazine to be filled from a charger, LoC 11753.

The arm is fitted to carry the sword-bayonet, pattern 1903, LoC 11716, 11717, on the nosecap. The barrel, sights, handguards and nosecap are identical with those described in LoC 11947.

Body— The body of the rifle is altered to receive the charger and the new locking and safety bolts.

Bolt— The bolt has the cover lugs and extension on the rear

end for safety catch arrangement removed, and striker hole bushed. In the M.L.M. Mark II rifles the cover lugs only have been removed.

Bolt head— The bolt head is altered to receive the charger guide, and has a slot cut in the screwed end of the bolt head which acts as a key when stripping and assembling the striker and cocking piece.

Trigger guard— The guard has a slot cut in the back to lighten it, and is freed at the front to clear the stop clip of the magazine.

Stock, fore-end— The stock fore-end has a liner fitted and glued in; the swivel slot and lower band pin hole are filled up, and the fore-end reduced externally; and an extra stop pin for the outer band is fitted.

Stock butt— The stock butt is reduced externally, bored longitudinally with four holes for lightness, the stock bolt hole deepened, and recess at socket end cut to clear locking bolt, and recess for butt plate strap filled up; it is also fitted with a marking disc. The butt plate is of sheet steel without a trap for the oil bottle, which is not carried in the butt.

When the stock butt is small at the socket end the shoulder and end are cut back 3/16 inch. This shortens the converted rifle as compared with the new by about 1/4 inch. In packing converted rifles an additional packing piece 1/4 inch thick will be required, inserted at the end of the chest behind the bridge fittings.

All dimensions, weights, &c., are the same as for the "rifle, short, M.L.E., Mark I", LoC 11947. All components are interchangeable with the new short rifle, M.L.E., LoC 11947, with the following exceptions—
    Screw, trigger guard, back
    Washers, rivet, fore-end (2)
    Spring, sight, aperture
    Screw, spring, sight, aperture
which are the same as for the rifle, M.L.E., Mark I*.

11950—Tin, mineral jelly, 3-oz. (Mark I)   L   29 Oct 1903
    Tin, with cork and spoon, Rifle, short, M.L.E.

Cover, tin, mineral jelly, 3-oz. (Mark I)   L   4 Dec 1903
    Canvas, with stud.

As materials for cleaning rifle will not be carried in the butt of the short rifle the above-mentioned articles have been sealed for carrying them, and will be manufactured in future. The use of the "Bottle, oil, Mark III" will be confined to the carbine and long rifle.

The tin is made with two compartments, the large one for mineral jelly, the small one for the folded pull-through. The former is closed by a cork to which a spoon and a brass lifting ring are fitted. The lid of the tin is shaped to receive strips of flannelette.

Mineral jelly (LoC 11957) for greasing the bore of the rifle, the feet, and boots will be carried in the large compartment.

The canvas cover in which the tin is carried is provided with a stud and stud hole to fasten the top flap. A loop at the back of the cover enables it to be carried on the waist belt by infantry, and on the bandolier by cavalry.

11951—Tools, armourers— 13 Oct 1903
  Anvil, stock butt, rifle, short, M.L.E.
  (Mark I) C
  Gunmetal, for attaching stocks butt.

A pattern of the above-mentioned anvil has been sealed to govern future manufacture.

The anvil is for use of armourers when attaching stocks butt; it is provided with a stud to enter the stock bolt hole, and the under surface is made concave to fit the convex surface of the butt. The butt will be inserted in the socket of the body, and the anvil placed in position on the end of the butt, which will then be driven home with a mallet or hammer. The anvil should then be withdrawn and the stock bolt inserted and screwed up, care being taken that the square end of the stock bolt protrudes through the face of the body about 1/8 inch, and is in correct vertical position.

11975—Cordite, T. C 1 Sep 1903
  Cordite, M.D.T. C

Such cordite or cordite, M.D., as may be in future made in tubular form will be described as shown above, the addition of the letter "T" indicating "tubular." The method of indicating the size will be as follows—

The mean external and internal diameter in hundredths of an inch of the finished cordite will be shown as the numerator, and the length of the sticks as the denominator of the fraction denoting size, e.g.—

"Cordite, M.D.T., size $\dfrac{50-17}{26}$

represents tubular M.D. cordite— external diameter, 0.5 inch, and internal diameter, 0.17 inch; length 26 inches.

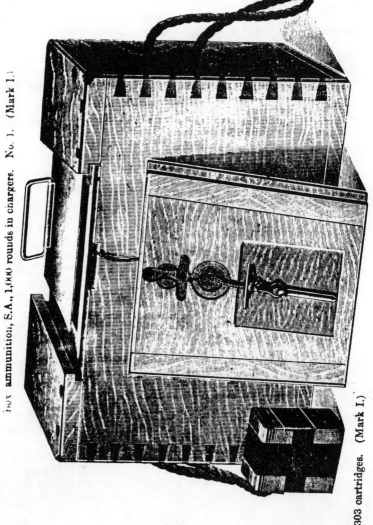

Box ammunition, S.A., 1,000 rounds in chargers. No. 1. (Mark I.)

Case, chargers, ·303 cartridges. (Mark I.)

11980—Box, ammunition, S.A., 1,000 rounds,   7 Jul 1903
   .303-inch, in chargers—   29 Sep 1903
    No. 1, Mark I   L   28 Oct 1903
    White wood, with tin lining.   20 May 1903
    No. 2, Mark I   L   15 Jun 1903
    Mahogany, with tin lining.   15 Jun 1903

   Case, chargers, .303-inch cartridges, Mark I L
   Cardboard; to hold 20 cartridges in
     four chargers.

  1. New patterns.

    Boxes, ammunition, S.A.—
    Wood; with tin lining.
    Home and special, Mark XIV   L
    G. S.   C

  2. New distinguishing labels when packed
     with ammunition in chargers.

    1. Patterns of the above-mentioned Nos. 1 and 2 boxes, and a drawing (C.I.W. 700) of the case, have been approved to govern manufacture. These boxes will eventually supersede, for land service, the "Boxes, ammunition, small-arm, home and special, Mark XIV," and the "G.S., Mark XI", respectively.

Box, No. 1.
    The box, which is for use in packing 1,000 rounds of .303-inch cartridges in chargers, is made to the form shown on the accompanying drawing. The box is of white wood, with a tin lining, and is provided with a sliding lid, which is secured by means of a half round brass split pin, having a T-shaped handle attached to it. The tin lining is provided with a closing plate fitted with a handle. Each end of the box is provided with a rope handle for lifting purposes.

Box, No. 2.
    This box is generally similar to the No. 1 box, but is made of mahogany or teak.

Case, chargers.
    The case is made of cardboard, glued and stapled together, to the form shown on the accompanying drawing. It is for use in packing 20 rounds of .303-inch cartridges held in four chargers.

    2. Boxes, A.S.A., home and special, Mark XIV; G.S., Mark XI: These boxes when packed as above with .303-inch ammunition in chargers, for land service, will contain 840 rounds, and when so packed will have distinguishing labels affixed to them which will clearly indicate the contents.

Dimensions, &c.

	Boxes, A.S.A., 1,000 rounds, Nos. 1 and 2.	Case, chargers.
	inches.	inches.
Length	16·5†	2·75
Width	8·312†	1·75
Depth	10·625†	3·375
Tonnage	ton ·02108	..
	No. 1.	No. 2.
	lb. oz.	lb. oz.
Weight { empty (about)	11 8	8 8	..
{ filled (about)	80 8	77 8‡	..

† Not to exceed for stowage.
‡ Boxes made of teak will weigh, filled, about 81 lb.

12057—Implement, action, rifle, short,　　　2 Feb 1904
　　M.L.E., Mark I.　　　　　　　　　　　　C
　　Steel, 3-blade; also sword-bayonet,
　　pattern 1903.

A pattern of the above-mentioned implement, as shown in the accompanying drawing, has been sealed to govern manufacture.

The implement is for use in stripping and re-assembling the stock fore-end and action, and testing the length and radius of striker point of "Rifles, short, M.L.E.," also for stripping and re-assembling the bolt of "Sword-bayonet, pattern 1903."

The following is a description of the various parts of the tool, showing the screws, &c., for which they are used.

　　(a) Driver, screw .. For screws— band, inner; cap, nose, back; guard, trigger, front; plate, butt; sight dial, fixing; spring, sight, aperture.
　　(b) Driver, screw .. For screws— bed, back sight, back and front; cut off; extractor; keeper, fine adjustment; handguard, cap; protector sight, top; nut, keeper, striker; and protector, sight, side.
　　(c) Driver, screw ... For screws— band, outer; swivels, piling and butt; cap, nose, front; disc, marking, butt; ejector; guard, back; sear; spring, sight, back; and stop, charger guide.

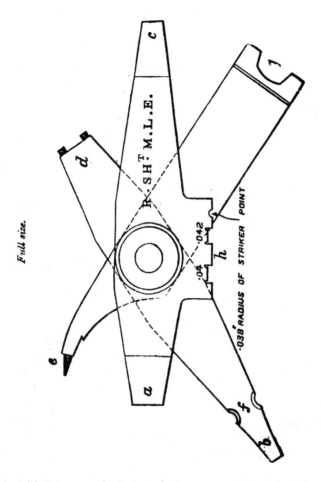

(d) Driver, screw, forked ... For screw, sight, dial, pivot.
(e) Tool, extractor spring, removing and replacing.
(f) Tool, expanding magazine link.
(g) Driver, screw, forked ... For nut, bolt, sword-bayonet, pattern 1903.
(h) Gauge, striker point, height and figure.

Instructions for the issue of this pattern "Implement, action," will be given with the orders for the supply of the short rifles.

12058—Sword, pioneers	C	28 Feb 1903
Scabbard, sword, pioneers	C	17 Nov 1903
Swords, buglers	L	1 Dec 1903
Scabbards, swords, buglers	L	

Obsolete.

The above-mentioned swords and scabbards, LoC 10974 & 7953 are hereby declared obsolete, and existing stocks will be dealt with as follows—

Naval Service

Swords and scabbards, pioneers— As notified in Admiralty letter dated 13th March, 1903, G 1403/03.

Land Service

Swords and scabbards, pioneers and buglers— Returned to store for transmission to Weedon for disposal.

12060—Sling, rifle, web, G.S., Mark I   L   13 Nov 1903
  Also R.A. carbine, 46 ins. by 1¼ ins.   1 Jan 1904

Alteration to fittings.

In future manufacture of the above-mentioned sling, LoC 10442, the brass catch or hook will be ½ inch longer and secured to the sling by two rivets instead of by eyelets.

The larger fittings will also be used for the repair of existing slings, when necessary, and the following stores should be demanded for this purpose—
  Catches, brass, large, for web slings   2
  Rivets, copper tinned, 3/8 inch, No. 12 W.G.   4

12073—Boxes,   14 Aug 1903
   Ammunition, S.A. and Q.F., Cartridge   20 Nov 1903
   Case, powder, 100 lb.

Red cross denoting varnished or painted linings to be abolished.

With reference to LoC 7455, 8285 and 9087— In future, the markings of a red cross on boxes and cases of the above-mentioned descriptions (to distinguish those which have varnished or painted linings) will be discontinued.

The red cross will be obliterated from all existing boxes and cases, locally. If, in carrying out this operation, any boxes or cases are found in store with the linings either unvarnished or unpainted, such packages should be issued first. The linings of any packages affected which are not yet varnished, will be varnished as opportunities offer.

12091–Rifle, short, magazine, Lee-Enfield   2 Nov 1903
   converted, Mark I   L   26 Feb 1904
   Converted from "Rifle, magazine,
   Lee-Metford, Mark I*"

Fitting of new pattern bolt head.

With reference to LoC 11948– An improved method of converting the bolt of the above-mentioned rifle has been approved. It consists in boring the front end of the bolt to receive a threaded bush, which is fixed by a portion of the original bolt-head screw, the bush and screw being brazed in. This method of conversion enables a bolt head common to all "Rifles, short, M.L.E." to be used.

The paragraphs regarding the bolt, and list of components special to the above-mentioned rifle, in LoC 11948, should now read as follows–

Bolt– The bolt has the cover stops and a portion of the top of the rib removed; the cover and bolt-head screw holes filled in. The following components will now be special to this rifle–
   † Body
   † Bolt
   † Stock, butt
   † Stock, fore end
† Components marked thus are converted from "Rifle, M.L.M. Mark I*".

12117–Box, ammunition, S.A., 600 rounds,   3 Jun 1903
   .303-inch, in chargers, Mark I   L   7 Dec 1903
   Mahogany, with tin lining; or 840 rounds
   in paper packets.

1. New Pattern.

   Boxes, ammunition, S.A.–
   780 rounds, Mark I   L
   Wood, with three tin boxes, each tin box
   containing 260 rounds, or 200 rounds
   in chargers, .303-inch.

   750 rounds, .303-inch, Mark I   L

2. No more to be provided.

1. A pattern (design R.C.D. 10408A) of the above-mentioned 600 rounds box has been sealed to govern future manufacture for Colonial services. The box is for use in packing either 600 rounds of .303-inch cartridges in chargers (LoC 11753), or 840 rounds in

ordinary paper packets.

It is made of mahogany, with a tin lining, and is provided with a sliding lid, which is secured by means of a half round brass split pin, having a T-shaped handle attached to it. The tin lining is provided with a handle. Each end of the box is provided with a rope handle for lifting purposes.

Dimensions, &c.
Length, over handles, not to exceed for stowage . . 12.375 ins.
Width, not to exceed for stowage . . . . 9.812 ins.
Depth, not to exceed for stowage . . . . 8.625 ins.
Tonnage . . . . .01515 ton
Weight, empty . . . . 11 lb. 5 oz.

2. A certain number of the above-mentioned 780 rounds boxes have recently been issued for colonial service, and a drawing (R.C.D. 10150A) has been sealed for record.

This box differs from the 600 rounds box in dimensions, in having a rope handle at one end only, and in being provided with three tin boxes, instead of a single tin lining. The tin boxes will each contain 260 rounds of .303-inch cartridges in ordinary paper packets or 200 rounds in chargers.

Consequent on the introduction of the 600 rounds box, no more "Boxes, ammunition, S.A., 780 rounds" or "750 rounds, .303-inch" (LoC 10750) will be made, and so soon as the existing stock is used up those patterns will be regarded as obsolete.

12129—Cut-off, rifle, short, M.L.E. (Mark I)   N   18 Feb 1904
    and rifle, short, M.L.E., converted, Mark II

Full size.

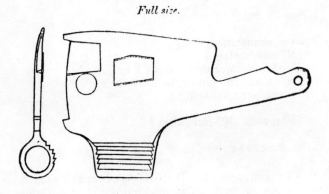

With reference to LoC 11850— A pattern of the above-mentioned component, as shown in the accompanying drawing, has been sealed in substitution of that described in above quoted paragraph. It differs in the following particulars—

The clearance for the magazine stop clip at front has been increased, and the lead for cartridge head increased at rear.

12155—Boxes, ammunition, small arm 27 Jan 1904
16 Feb 1904

Labels to distinguish boxes containing .303-inch cartridges in chargers.

In continuation of LoC 7603— In order to distinguish boxes containing .303-inch cartridges packed in chargers from those with cartridges in ordinary paper packets, the labels on the various boxes affected will be as illustrated in the annexed "typical" drawings.

Typical Distinguishing Labels

Scale ½

Printing in red; with "CHARGERS" in black across lower square.

12156—Boxes, ammunition, S.A.—   31 Aug 1903
    Pistol, Enfield, 240 rounds   L   1 Dec 1902
    .303-inch, half, naval   N

New cross handle fastening.

The sliding lids of boxes of the above-mentioned description will, in future manufacture, be secured by means of a half round brass split pin having a T-shaped handle attached to it.

12157—Boxes, ammunition, S.A., 1,000 rounds,   12 Jan 1904
    .303-inch, in chargers—
    No. 1, Mark I   L
    No. 2, Mark I   L

Altered dimensions and weights.

With reference to LoC 11980— The dimensions and weights of the above-mentioned boxes should be as shown below, instead of as therein stated.

Dimensions, &c.

	Boxes, A.S.A., 1,000 rounds. Nos. 1 and 2.		
Length ⎱ Width ⎬ Not to exceed for stowage Depth ⎰	inches. 17 8·312 10·85		
Tonnage	ton. ·02218		
		No. 1 box.	No. 2 box.
		lb. oz.	lb. oz.
Weight { empty .. .. about		11 10	14 2
{ filled .. .. about		80 10	83 2

12184—Rifle, magazine, Lee-Metford (Mark II)   C   16 Mar 1904

Fitted with barrels having Enfield rifling with body attached. Nomenclature when so fitted.

With reference to LoC 11498— "Rifles, magazine, Lee-Metford Mark II" (with Mark II breech bolt), will still retain their original nomenclature, viz., "Mark II" notwithstanding the fact that they may have been fitted with Enfield barrels with body attached.

In order to make this clear, such bodies will be stamped "II.", and the letters "L.E. I." crossed out, for which purpose the necessary tools will be supplied on demand by the Chief Ordnance Officer, Weedon.

Components for these rifles will be demanded as for "Rifles, M.L.M., Mark II" and officers commanding units having any of these rifles on charge should provide for a proportion of Mark II spare components (such as strikers and cocking pieces) for the same.

Rifles which have been fitted with Enfield barrels with or without bodies, and which have an M.L.E. breech bolt, are M.L.E. rifles, and should be marked on the body "L.E. I." It must be distinctly understood that the pattern of the breech bolt determines the nomenclature of the above-mentioned rifles, and care should be taken to ensure that the marking on the body, the pattern of the breech bolt, and the spare components to be demanded correspond in each case.

It may be found possible to largely secure the required uniformity in pattern by interchanging, regimentally, the breech bolts where the difference at present exists, and that course should be followed wherever possible before re-marking the bodies of the rifles.

12185—Scabbard, sword-bayonet, pattern 1903, 11 Mar 1904
naval. (Mark II) N
Brown leather; also sword-bayonet, pattern 1888.

A pattern of the above-mentioned scabbard, a drawing of which is attached, has been sealed to govern future manufacture.

This scabbard is generally similar to the pattern described in LoC 11811, but rivets are used instead of stitches for the loop. The scabbard is interchangeable on sword-bayonets, patterns 1888 and 1903.

Length of scabbard over all . . . . .16 5/8 ins.
Weight . . . . .7½ oz.

(For illustration, see following page).

12186—Tin, mineral jelly, 3-oz. (Mark I) C 29 Oct 1903
Tin; with cork and spoon ; Rifle, short, M.L.E. 4 Dec 1903
  22 Apr 1904
Cover, tin, mineral jelly, 3-oz. (Mark I) C
Canvas, with stud.

With reference to LoC 11950— It has been decided to extend the use of the above-mentioned articles to the Naval Service. The

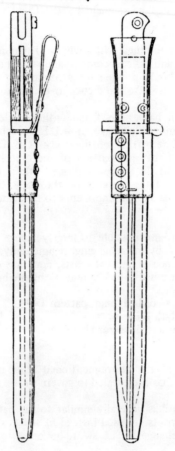

Scale ¼.

LoC 12185.

distinguishing letter has, therefore, been altered from "L" to "C".

12220–Action, skeleton, rifle, short,   27 Feb 1904
    M.L.E. (Mark I)   C   11 May 1904
    Also Rifles, short, M.L.E., converted,
      Marks I and II.

    A pattern of the above-mentioned article has been sealed to govern future manufacture.

The action of the "Rifle, short, M.L.E., converted, Mark I," differs slightly in detail, and the components are not interchangeable with "Rifles, short, M.L.E., Mark I; and converted, Mark II," but the principle of construction being the same, this pattern of skeleton action adequately illustrates the mechanism of any pattern of short rifle.

Skeleton actions for Naval Service differ only from those for Land Service in being fitted with a cut-off.

12221 – Action, skeleton, rifle, short, M.L.E.   18 Mar 1904
    converted (Mark I)   C   11 May 1904
    Converted from Action, skeleton, rifle, M.L.E.

A pattern of the above-mentioned article has been sealed to govern conversions as they may be required.

The conversion consists in altering the body, bolt, guard, and stock-butt, and fitting the action with the short rifle components. The body differs slightly from the "Action, skeleton, rifle, short, M.L.E. (Mark I.)" described in LoC 12220 in the following manner—
    The body is converted, and the clearance on left side of body for the thumb, when loading with the charger, is shallower.

In all other particulars the converted skeleton action is similar to the new. Skeleton actions for Naval Service differ only from those for Land Service in being fitted with a cut-off.

12223 – Belt, waist, brown, Warrant Officers   8 Mar 1904
    (Mark I)   L
    and S.S. dismounted services, with universal locket and detachable slings.

    1. New pattern.

    Belts—
      Pouch, buff—
        G.S., Mark II   L
        With 2-inch buckle and stud; W.O. and S.S. dismounted services.
        G.S., Mark II*   L
        With 2-inch buckle and stud; converted from enamelled G.S., W.O., and S.S. dismounted services.
      Waist, black—
        Warrant Officers (Mark II)   L
        and S.S. rifles, japanned, with snake hook.
      Waist, buff—
        Warrant Officers—
        and S.S. dismounted services.

Artillery   L
Also Engineers, with plate.
G.S.   L
With universal locket.

Pouches, writing materials—
Morocco leather.
Artillery, Mark II   L
With gun ornament.
Conductors, Mark II   L
With E.R. ornament.
Engineers, Mark II   L
W.O. and S.S., with Royal Arms ornament.

2. To become obsolete.

1. A pattern of the above-mentioned "Belt, waist, brown," has been sealed to govern future manufacture for all dismounted services except Foot Guards.

It differs from the previous pattern (LoC 6205) in being made of brown leather instead of buff, and in the slings being detachable, for removal on active service.

2. The "Belt, pouch, buff, G.S." (LoC 62050 and "Pouch, writing materials, Artillery, Conductors, and Engineers" (LoC 11037) will be worn only with the "Belt, waist, buff" (LoC 6205) and when existing stock has been used up these articles will become obsolete, and also the "Belt, waist, black, W.O., Mark II" (LoC 6475).

12292—Boxes—   9 Jul 1904
    Ammunition, S.A. & Q.F.
    Cartridge.

    Case, powder, 100-lb.

1. Red cross denoting varnished or painted linings to be abolished.
2. Unvarnished or unpainted tin linings to be varnished locally, before issue, as opportunities offer.
3. LoC 12073 cancelled.

With reference to LoC 7455, 8285, 9087, 9622—
1. In future, the marking of a red cross on such of the above-mentioned packages as are provided with tin linings (to distinguish those which have varnished or painted linings) will be discontinued.

Existing boxes or cases having the red cross will be dealt with, locally, as follows—

The stock will have the red cross obliterated by being barred through with black paint as opportunities offer. Sufficient boxes will be done to ensure that none are issued with the cross on them.

2. If, in carrying out the above operation, any packages are found in store with their tin linings neither varnished nor painted, the opportunity should be taken to varnish such tin linings before issue. Packages having the zinc linings should not be varnished, vide LoC 9622.

3. LoC 12073 is hereby cancelled.

12326–Case, 200 scabbards, sword-bayonet,   21 Jun 1904
    pattern 1903 (Mark I)   C
    Wood, with fittings; also for 220 scabbards, sword-bayonet, pattern 1888.

Use extended to Naval Service.

With reference to LoC 11849– The use of the case therein described having been extended to Naval Service, the nomenclature and distinguishing letter have been altered as shown above.

12327–Cover, tin, mineral jelly, 3-oz. (Mark II)   C   8 Jul 1904
    Canvas, with stud.

With reference to LoC 11950 and 12186– A pattern of the above-mentioned article has been sealed to govern future manufacture. It differs from the Mark I pattern, described in the above-quoted paragraphs in the following manner–
    The back and bottom are formed in one piece and the loop is made broader.

12328–Sword, practice, gymnasia, pattern 1904   25 Mar 1904
    (Mark I)   L
    Without scabbard.

A pattern of the above-mentioned sword has been sealed to govern future manufacture.

It differs from the pattern 1899 (LoC 9984) in being slightly stiffer in the blade, and the hilt is not perforated. With a weight of 2½ lb. placed in the guard, the blade should recover straightness, and with a weight of 4 lb. placed in the guard, the blade should give way.

12389–Bandolier equipment, pattern 1903–   20 Oct 1903
    Bandolier, 50 rounds (Mark I)   L
    Brown leather, with five pockets and steadying strap, two buckles and runner.

Carrier, greatcoat.

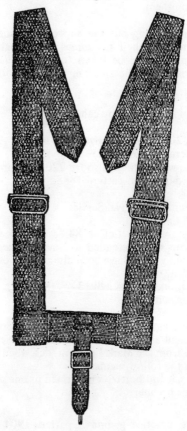

Cover, mess tin.

Belt, waist (Mark I)   L
Brown leather, with buckle and runner; also for Cavalry pistol.

Pockets, cartridge—
Brown leather
    15 rounds (Mark I)   L
    10 rounds (Mark I)   L

Strap, greatcoat (Mark I)   L
Brown leather, 37 inches by ¾ inch, with double roller buckle.

Carrier, greatcoat (Mark I)   L
Web.

Cover, mess tin, dismounted men (Mark II)   L
Canvas, with loops for waistbelt.

Patterns of the above-mentioned articles have been sealed to govern future manufacture.

Bandolier— The bandolier is cut on a curve in order to fit closely to the wearer's shoulder, and has five pockets rivetted on the front, each of which is designed to take 10 cartridges in two chargers. A small strap is fitted inside each pocket to secure the front charger after the back one has been removed (see drawing).

Belts, waist— The waistbelt is a plain strap with a buckle and runner at one end and holes punched at the other, to enable it to be adjusted to the man's waist (see drawing).

Pockets, cartridge— These are for use on the waistbelt. The 10-round pockets are similar to those attached to the bandolier, except that they are provided with a leather loop on the back for sliding them on to the waistbelt.

The 15-round cartridge pocket is rectangular in shape. It has a loop at the back for sliding on to the waistbelt, but has no securing strap inside. The three chargers are placed in the pocket in sandwich fashion, the centre one having the bullets point down. A brass dee is fitted to the back of the pocket to take the hook on the shoulder straps of the greatcoat carrier (see drawing).

Strap, greatcoat— The "Strap, greatcoat" is similar to the 1888 pattern, but is made of brown leather (see drawing).

Carrier, greatcoat— The "Carrier, greatcoat" is made of cotton web of the colour of the Service dress, and is fitted with brass loops,

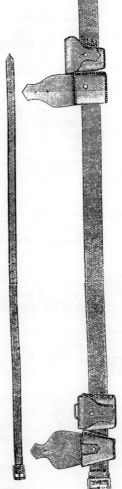

Bandolier, 50 rounds. Strap, greatcoat. Belt, waist, with Pockets, cartridge.

Haversack, G.S. (Mark I.)

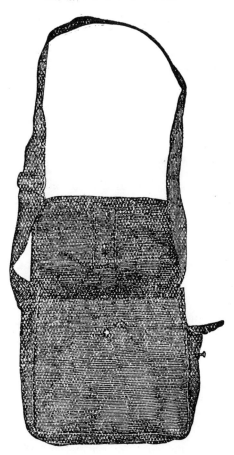

slides and clips. The shoulder straps of the carrier are fitted with brass hooks for hooking into the dees on the 15-round pockets on waistbelt. The short steadying strap at back has a brass hook for hooking into a ring on the cover of the mess tin (see drawing).

Cover, mess tin— The "Cover, mess tin" is made of flax canvas of the colour of the Service dress. It is fitted with two leather loops at back for sliding on to waistbelt. A brass ring is attached to the top of the cover at the back into which the steadying strap of the greatcoat carrier is hooked. The cover is fastened by a button.

The Mark I "Cover, mess tin" was not published in the List of Changes.

The waistbelts, greatcoat straps, and mess-tin straps of the valise equipment, patterns 1882 and 1888, will be used up, and the remaining articles of these equipments now in use withdrawn as issues of the new articles are made.

A brown leather waistbelt, and greatcoat strap, will be issued when the existing waistbelts, greatcoats and mess-tin straps, are worn out.

Haversack, G.S. (Mark I)   L   19 Aug 1903
Drab canvas; shoulder strap with loop and slide.

A pattern of the above-mentioned article (see drawing) has been sealed to govern future manufacture.

The haversack is made of drab-coloured canvas, and is fitted with a cotton web shoulder strap of the same shade of colour; the strap is 53½ inches in length and 2 inches in width, and is fitted with a brass loop and slide. A small pocket with flap is attached to one side for the "emergency ration."

12421—Aim corrector (Mark II)   C   22 Jul 1904
    All .303-inch and .45-inch rifles and carbines.   15 Sep 1904

A pattern of the above-mentioned aim corrector has been sealed to govern future manufacture.

It differs from the Mark I described in LoC 6628, in the following manner—
The bow is made in the form of a spring clip, to fit over the handguard, barrel and fore-end of the arm. The box is arranged so that the glass, the tint of which is deeper than that of the Mark I, may be placed in two positions at an angle of 45 degrees with the axis of the barrel, thus enabling the aim corrector to be used, and the instructor to stand on the right or left of the arm when aim is being taken. The aim corrector can be used with either long or short rifles.

"Aim correctors, Mark I" will become obsolete as soon as the stock is used up.

12422—Tools, armourers   3 Aug 1904

Patterns of the undermentioned tools have been sealed to govern manufacture for use with the Rifle, short, M.L.E.—

    Tools, armourers—
       Tool, sight line   C Steel; reversible
          Rifle, short, M.L.E. (Mark I)

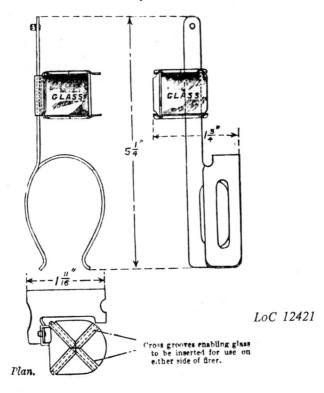

LoC 12421

Scriber, tool, sight line Rifle, short, M.L.E. (Mark I)	C	Steel
Tool, removing plate, keeper screw windgauge Rifle, short, M.L.E. (Mark I)	C	Wire, steel, circular handle
Drift, pin fixing washer, pin axis, sight back Rifle, short, M.L.E. (Mark I)	C	Wire, steel
Drift, pin fixing stud, head catch, slide, sight back Rifle, short, M.L.E. (Mark I)	C	Wire, steel
Driver, screw, nut, bolt Sword-bayonet, pattern 1903 (Mark I)	C	Steel, forked with wood handle

The "Tool, sight line" and "Scriber, tool, sight line" are for use of armourers when re-lining the windgauge and fine adjustment of the back sight of Rifles, short, M.L.E. The tool is reversible, so that the lining may be done from both sides of the leaf, to ensure the line being central. When re-lining with the above tools after re-blueing, particular care should be taken that the deep machine cut lines on wind-gauge and fine adjustment are opposite each other.

To use the "Tool, removing, keeper plate, wind-gauge screw," remove fine adjustment keeper screw, and fine adjustment with wind-gauge, insert the tool in extracting hole of the keeper plate, and lift out.

12423—Accoutrements. 4 Jul 1904

1. New patterns.
2. Nomenclature.
3. Obsolete.

1. Patterns of the undermentioned articles have been sealed to govern future manufacture for all services, except Foot Guards—

Belts, shoulder, brown—		Leather
Mountain Artillery—		
Dismounted men (Mark I)	L	With attached frog
Mounted men (Mark I)	L	With stud attachment for belt
Belts, waist, brown—		Leather
Mountain Artillery (Mark I)	L	Mounted men, with plate and detachable frog
Loop, brown, dirk (Mark I)	L	Leather, Highland regiments

They are generally identical with the buff equipment, with the following exception—
Belt, waist, brown, Mountain Artillery.

This differs from the previous pattern (LoC 8370) in being made of one piece, with two dees riveted on to take straps of frog, instead of in three pieces with two brass rings.

2. Consequent on the introduction of articles in brown leather, the nomenclature on page 83 of Priced Vocabulary of Stores, 1902, has been amended to read as now shown—

Frog, brown, sword	L	Leather, 7¾ ins., with loop, R.A.M.C.
Frog, buff, sword	L	Leather, 7¾ ins., with loop, R.A.M.C.

Knot, sword, brown, G.S.   L   Flat, with acorn.

The nomenclature of "Frog, brown, sword, pioneers" (LoC 4855) and "Frog, buff, sword" (LoC 10258) has also been amended to read as shown.

The "Knot, sword, brown, Infantry" (LoC 4855) will be re-sealed as now described, and be issued for all services, except Household Cavalry.

3. The following articles will become obsolete—

Belt, waist, black, transport serjeants, rifle battalions.	L	Leather
Belts, waist, buff—		
Cadets—		
Sandhurst—		
With locket	L	
Without locket	L	
Woolwich	L	With plate.
Mountain Artillery	L	Mounted men, with plate and detachable frog.
Under Officers—		
Sandhurst	L	With locket
Woolwich	L	With plate
Frogs, black, sword	L	7¾ ins., with loop, pioneers
Knot, sword, black	L	Leather, with runner
Knot, sword, black, transport serjeants, rifle battalions	L	Leather, flat, with runner and brass button
Knot, sword, brown—		Leather, with runner
Military Mounted Police—		
S.S.	L	Plaited, with acorn
O.R.	L	Flat, with brass button
Knot, sword, buff—		With runner
Cavalry, line, O.R.	L	and all ranks, H. & F. Artillery and S.S.
Warrant officers	L	
Loop, buff, drum carriage	L	With buckle for O.P. carriage
Pouch, ammunition, black, Artillery, Mark II	L	Leather, 20 rounds, gun ornament, limber gunners.

Existing stocks will be used up with the exception of the two last named, which being no longer required will be returned to store.

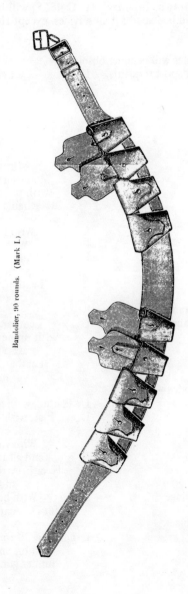

Bandolier, 90 rounds. (Mark I.)

LoC 12424

12424—Bandolier, 90 rounds (Mark I)   L   29 Sep 1903
  Brown leather, with nine pockets,
  double buckle and runner.

A pattern of the above-mentioned bandolier has been sealed to govern future manufacture for mounted services.

It is made of brown leather, and is shaped as shown in the accompanying drawing. The pockets are made to carry 10 rounds of ammunition in two chargers. A small guard strap to prevent the second charger from falling out when the first has been withdrawn, is fitted to each pocket as in LoC 12389. No steadying strap is fitted to this bandolier.

12425—Belt, waist, brown—   27 Aug 1904
  Leather.
   Cavalry—
    S. S. (Mark I)   L
    Gilt furniture with snake hook, all W.O.
    and S. S. carrying Cavalry swords.
    O. R. (Mark I)   L
    Brass furniture with snake hook, all
    other mounted men carrying Cavalry
    swords.

1. New patterns.

   Belts, waist, brown—
   Leather.
    Military Mounted Police   L
    Pistol, Cavalry   L

   Belts, waist, buff—
    Cavalry, pattern 1885—
    S. S.   L
    Gilt furniture with snake hook, all W.O.
    and S. S. carrying Cavalry swords.
    O. R.   L
    Brass furniture with snake hook, all
    other mounted men carrying Cavalry
    swords.

2. To become obsolete.

1. Patterns of the above-mentioned belts have been sealed to govern future manufacture for all mounted services, except Household Cavalry. They are made similar to the Mark II buff belts mentioned in LoC 5377, as modified by LoC 10765.

2. No more supplies of the above-mentioned belts (LoC 3597, 5377, 7496 and 10765) will be obtained, and when present stock has been used up they will become obsolete.

12470—Cuirass— 27 Sep 1904
    Steel, with studs, washers, lining and edging.
        Back (Mark II)   L
        Front (Mark II)   L
    Straps—
        Shoulder (2) (Mark II)   L
        Leather, with plates, scales and studs.
        Waist (Mark II)   L
        Buff leather, in two parts, with buckle.

Patterns of the above-mentioned articles have been sealed to govern future manufacture.

The "Cuirass, back and front" and "Straps" differ from the Mark I pattern (LoC 2927) in small manufacturing details only.

12471—Bandolier equipment, pattern 1903—   26 Oct 1904
    Carrier, greatcoat (Mark II)   L   3 Nov 1904
    Web.

A pattern of the above-mentioned article has been sealed to govern future manufacture.

It differs from the previous pattern, Mark I (LoC 12389), in the following particulars— The clips and hooks are of a different design, and the slides are narrower. The loops on the connecting piece are wider to allow the slides to pass through more freely for adjustment. The lead of the brace round the coat is reversed, the web on the double being between the man's back and the coat.

12526—Bandolier equipment, pattern 1903—   22 Nov 1904
    Cover, mess-tin, dismounted men
    (Mark II)   L
    Canvas, with loops for waist-belt.

Alteration.

With reference to LoC 12389— A slight alteration has been approved for future manufacture of the "Cover, mess-tin" by dispensing with the two canvas loops for mess-tin strap, the strap being no longer used for this service.

12549—Cartridges, S.A. dummy, drill,   7 Sep 1904
    .303-inch rifles or carbines   18 Nov 1904
    (Marks I, II and III).   C

1. To have holes bored through case.
2. Retinning of Marks I and II.
3. Tinning of case for Mark III to be discontinued.

1. Dummy drill cartridges for .303-inch arms (LoC 6057, 9519, 11832, 12451) will in future manufacture have holes bored in the case to further distinguish them from ball cartridges.

Those existing in land service will also have holes drilled in the case locally as follows—

Drill, in accordance with the following instructions, two holes 3/16 inch in diameter through the case on two diameters at right angles to each other—

Mark the position of the centre of the hole, and drill a small hole by means of a brace of any convenient size; open out the hole with a round file, and finish it with a rimer of the required size. The holes near the base are to be just above, and clear of, the cap chamber, the upper holes being midway between the bottom holes and the neck of the case.

2. When considered necessary, owing to the tinning having worn off, Marks I and II cartridges will be retinned, locally, by being immersed in the following boiling solution (the cartridges being previously cleaned by dipping in a weak solution of hydrochloric acid)—

Grain, tin . . . . .3 lb.
Bitartrate of potash . . . . .3 oz.
Alum . . . . .3 oz.
Common salt . . . . .3 oz.
Water . . . . .3 gallons.

The strength of the solution should be maintained by the addition of small quantities of the ingredients, as required, from time to time. Care must be taken, in drying, that the cartridges are thoroughly drained.

3. The case of the Mark III cartridge will not be tinned in future, and existing stocks of this "Mark" need not be retinned when the original tinning wears off. This cartridge is easily distinguished from ball cartridge by its wooden bullet.

12567—Case, regimental, 10 rifles, M.L.M.   11 Oct 1904
   3 Nov 1904

No more to be provided.

Case, 10 rifles, M.L.M.

To be converted for regimental use locally as required.

No more of the above-mentioned regimental cases will be made, the ordinary "Case, 10 rifles," fitted with lock and hinges being used in lieu.

Existing "Cases, 10 rifles" (ordinary), will be converted locally, as may be required, for use as regimental cases in accordance with the instructions detailed below, for which purpose the undermentioned stores will be supplied to officers concerned on demand.

Instructions.
(1) Hang the lid to the case by the two hinges supplied, each being at a distance of 5 inches from the end.
(2) Fit the case with a lock and catch at a distance of 16½ inches from the right-hand end of case.
(3) Fix to the back of case a backstop for lid ½ inch down from top edge.
(4) Paint the case on the outside.

Stores required for one case.

No.

Hinge— with 7 screws, iron, flathead, 1 inch long, gauge No. 16, and 2 countersunk head steel rivets, 5/16 inch diameter, cut 1½ inches long 2

Lock, 3½ inch, complete— with 7 screws, iron, flathead, 1 inch long, gauge No. 8 1

Backstop, yellow deal, 52 inches long, 7/8 inch thick by 1½ inches wide— with 6 screws, iron, flathead, 1¾ inches long, gauge No. 14 1

12568—Scabbard, sword-bayonet, pattern 1903,   9 Nov 1904
   Naval (Mark III)   N   2 Dec 1904
   Brown leather; also sword-bayonet, pattern 1888.

A pattern of the above-mentioned scabbard has been sealed to govern future manufacture.

It differs from the scabbard described in LoC 12185 in the following particulars—
The loop of the tag is much longer, and stitched instead of riveted, and the tag is stitched at the side of the locket instead of riveted.

Length of scabbard over all . . . . 16 3/8 - 16 5/8 ins
Weight of scabbard . . . . .7 ozs.

The scabbard is interchangeable on sword-bayonets, patterns 1888 and 1903.

12597—Cartridges, S.A. dummy, drill, .303-inch rifles or carbines. (Marks I, II and III)   C   7 Sep 1904
18 Nov 1904
14 Jan 1905

1. To have holes bored through the case.
2. Re-tinning of Marks I and II in Land Service.
3. Tinning of case for Mark III to be discontinued.
4. LoC 12549 cancelled.

1. Dummy drill cartridges for .303-inch arms will, in future manufacture, have holes bored in the case to further distinguish them from ball cartridges.

Existing Marks I, II and III dummy cartridges (LoC 6057, 9519, 11832 and 12451) will have two holes, 3/16 inch diameter, drilled through the case on two diameters at right angles to each other, in accordance with the following instructions. This service will be carried out in—
    Land Service.....................Locally.
    Naval Service ....Locally, and on board ships.

Instructions.
Mark the position of the centre of the hole, and drill a small hole by means of a brace of any convenient size; open out the hole with a round file, and finish it with a rimer of the required size. The holes near the base are to be just above, and clear of, the cap chamber, the upper holes being midway between the bottom holes and the neck of the case.

2. When considered necessary, owing to the tinning having worn off, Marks I and II cartridges in Land Service only, will be re-tinned locally, by being immersed in the following boiling solution (the cartridges being previously cleaned by dipping in a weak solution of hydrochloric acid)—
    Grain, tin.......................3 lb.
    Bitartrate of potash................3 oz.
    Alum ....3 oz.
    Common salt ....3 oz.
    Water ....3 gallons.

The strength of the solution should be maintained by the addition of small quantities of the ingredients, as required, from time to time. Care must be taken, in drying, that the cartridges are thoroughly drained.

3. The case of the Mark III cartridge will not be tinned in future, and existing stocks of this "Mark" need not be re-tinned when the original tinning wears off. This cartridge is easily distinguished from ball cartridge by its wooden bullet.

4. LoC 12549 is hereby cancelled.

12603—Flannelette (Mark III) 1 Sep 1904
    4 inches wide.

    A pattern of the above-mentioned flannelette has been sealed to govern future supplies.

    It differs from the previous pattern (LoC 6673) in width, being 4 inches wide, with transverse red stripes 2 inches apart.

12610—Pull-through, double (Mark I) L 22 Oct 1904
    All rifles and carbines.

    The "Pull-through, double (Mark I)", LoC 9320, 9774, is no longer to be used for cleaning rifles and carbines, except by armourers.

    All double pull-throughs issued to troops for this purpose will be withdrawn and returned to C.O.O., Weedon, except those in use for machine guns.

12611—Rifle, short, M.L.E. (New and converted) 25 Nov. 1904
    Fitting of spare bolt heads and charger guides.

    Spare charger guides for the repair of above-mentioned rifles in Land and Naval Service will, in future, be issued soft and fuller at the face against which the charger bears when loading.

    When fitting a new charger guide, the armourer will adjust this face by filing until the charger fits, care being taken to retain the original angle. He will then case-harden the charger guide, and adjust it to the bolt head.

    Spare bolts will be issued hardened as at present, but left longer at the front; they will be fitted to the rifle in accordance with the following instructions—

    Assemble the bolt head to bolt, insert in body, and test with .064-inch gauge. Should the bolt not close on the gauge, remove the bolt head from bolt, and having placed a piece of emery cloth (No. F) on a flat surface, rub the face of the bolt head on the emery cloth, maintaining a circular motion in order to preserve a flat surface, until sufficient metal has been removed to enable the assembled bolt to close over the .064-inch gauge, and not close over the .072-inch gauge.

    The face of the bolt head should now be tested with a file to see if it is hard. If found to be soft, the face of the bolt head should be re-hardened in the following manner—

    Wrap a piece of wet rag (old sheeting) four or five times round the tang of the bolt head; grip the tang, with the rag round it, in the tongs, and heat face of bolt head to a blood red in the forge. Then sprinkle over the face powdered yellow prussiate of potash, which should be allowed to remain on the face for about 10 seconds. Then dip in cold water; thoroughly clean, and adjust.

To enable armourers to carry out this re-hardening, a biennial allowance of yellow prussiate of potash will be allowed as follows—

800 arms.........................24 oz.
600 arms.........................18 oz.
500 arms.........................15 oz.

12612—Scabbard, brown leather, sword-bayonet,   3 Mar 1904
       pattern 1888, Land (Mark II*)   L

With reference to LoC 11151— A pattern of the above-mentioned scabbard (as shown in the drawing overleaf) converted from "Scabbard, Mark II" has been sealed to govern conversions as may be ordered.

The conversion consists in attaching a brown leather loop to the tag by means of two "Rivets, copper, tinned, with washers, 3/8 inch, No. 9." The position of the loop is such that the stud holes at the ends of the loop coincide with the stud hole in the tag. When the loop is fixed the edges of the tag are cut flush with the edges of the loop.

12697—Accoutrements, naval, pattern 1901—   20 Dec 1904
       Pockets, cartridge, rifle (Mark I)   N
       Brown leather, 45 rounds.

A pattern of the above-mentioned article has been sealed to govern future manufacture.

There are three 15-round pockets, similar to those mentioned in LoC 12389, but of leather, the same shade of colour as the naval accoutrements described in LoC 11110, riveted to a band of leather forming an elongated loop for attachment to waistbelt. These articles will supersede the "Bags, ammunition" referred to in LoC 11110, as the stock of the latter is used up.

The loop on the "Belt, waist" (LoC 11110) has been made removeable, in order that the pockets, when worn may not be forced back on the hip; this alteration can be carried out locally when required.

12714—Saddlery, universal—   1 Mar 1905
       Bucket, rifle, Cavalry (Mark I)
       Leather, reversible.

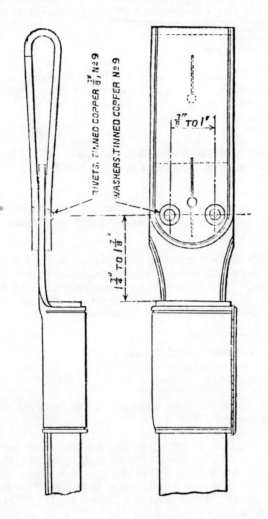

LoC 12612

LoC 12714 cont'd —

A pattern of the above-mentioned bucket has been sealed to govern manufacture.

It differs from the carbine bucket (LoC 8372) in being larger and shaped to take the short rifle. It is of stout bridle leather; the

A

Saddlery, universal—
    Bucket, rifle, Cavalry. (Mark I.) ... ... | Leather, reversible.

See § 12714.

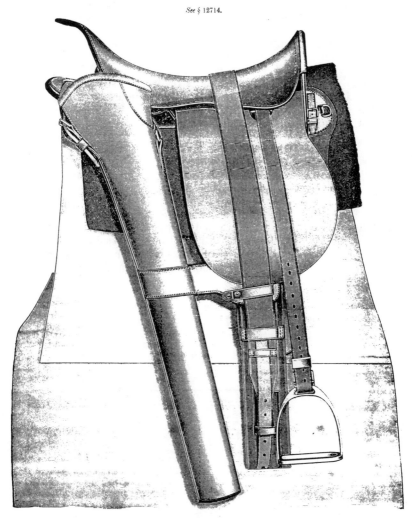

carbine bucket was of crop shoe leather. The front of the mouth of bucket, and the arm, are stiffened with steel plates. The arm is set on obliquely to place the bucket, when attached to the saddle, so that the butt of rifle is in rear of the rider's elbow.

The bucket is to be suspended from the saddle by passing the strap round the hind arch strut of steel arch saddles, and then through the brass link on the side bar cap (see following), the surcingle being passed through the loop of arm (see drawing A).

On Yeomanry pattern saddles the suspending strap should be passed downwards through the front dee and half twisted, and then downwards through the rear dee and buckled. The half twist in the strap, caused by this method, saves the wear of strap and protects the side of saddle. The link will not be required on the pannels of this saddle. Although the bucket is reversible, it will, as a general rule, be worn on the off side of the saddle.

Weight of bucket.....................2 lb. 8 oz.

Instructions for fitting brass links to numnah pannels and side-bar caps.

To allow the bucket to be attached as described above, the following alterations will be carried out by regimental artificers—
A brass link and leather chape are to be attached to the numnah pannel leather pockets, and, on such saddles as are at present worn without numnah pannels, on the side-bar caps.

The chape is to be sewn on with the rear edges 2 inches from the end of cap or pocket, as shown (see drawing B) and the pockets or caps secured by two brass screws.

B

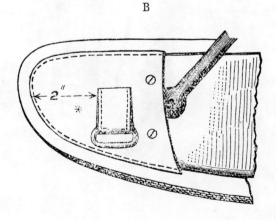

The undermentioned articles will be demanded by commanding officers concerned for the purpose of carrying out these orders, viz.-

Links, brass, 1 1/8 ins. by ½ in.	2 per saddle in possession.
Chapes, leather, 2¾ ins. by 1 1/8 ins.	2 per saddle in possession.
Thread, flax, fine	4 oz. per 50 saddles.
Screws, brass, ½ in., G9	4 per saddle.
Wax, black	1 oz. per 50 saddles.

* *On pockets and caps to be fitted to bars under 23½ inches in length, viz., 0 size, and those shortened to pattern 1902, the chape should be attached 1½ inches from the end of the cap or pocket.*

12741—Pistol, signal, Very's cartridge (Mark II)   C   15 Feb 1905
    Gunmetal

A pattern of the above-mentioned pistol has been sealed to govern future manufacture. It differs from that described in LoC 5181 in the following particulars—

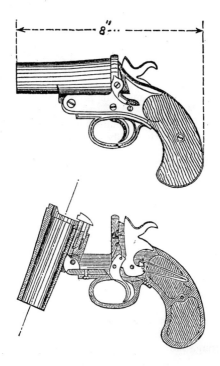

113

The barrel is strengthened. The extraction is positive, and partly withdraws the case from the chamber automatically on dropping the barrel. The hammer is arranged to rebound, and is fitted with a swivel for the mainspring. A trigger guard is provided, and a barrel catch fitted; the barrel catch prevents the hammer point striking the cap and firing the cartridge, until the barrel is securely locked.

Weight of pistol. . . . . .2 lb. 7 oz.
Length, over all. . . . . .8 inches.

12763—Case, regimental, 10 rifles, short,   3 Nov 1904
    M.L.E., converted   **L**
    Wood; with fittings; converted from "Case, regimental, 10 rifles, M.L.M." for spare arms, including side arms.

Conversion from "Case, regimental, 10 rifles, M.L.M."

With reference to LoC 12567— The "Cases, regimental, 10 rifles, M.L.M.," therein mentioned, will be converted and used for storing "Rifles, short, M.L.E."

The cases in possession of units equipped with short rifles will be altered locally in accordance with instructional print, R.C.D. 1167, which will be supplied to officers concerned on demand, the necessary material being obtained locally. A sample will, if necessary, be supplied to guide the local conversion of cases.

12764—Musket, fencing, short (Mark I)   **L**   14 Nov 1904
    With spring bayonet and indiarubber pad.
    Spanner (Mark I)   **L**
    Steel; removing and replacing barrel bush and bayonet button, Musket, fencing, short.

Patterns of the above-mentioned articles have been sealed to govern future manufacture.

The existing stock of O.P. muskets will be used up. The new pattern musket in length, weight and balance is similar to the "Rifle, short, M.L.E., with sword-bayonet". It differs from the "Musket, Mark V" described in LoC 8369, 9982, in the following particulars—
    The barrel is a plain tube, without studs, tapped at the front end to receive the barrel bush which is removeable. The barrel bush is screwed into the barrel by the milled head, across which flats are cut for the spanner.

    The hand-guard is in two pieces, top and bottom, and is held in position by two screws, one on either side; a third screw passes through the top portion of the hand-guard, barrel, and stock, and is screwed into the bottom portion of the handguard, locking the whole together.

The stock, butt, is provided with a pistol grip, similar to the "Rifle, short, M.L.E."

There are no straps fitted to the stock. The bayonet spring is 4 inches shorter, with 68 coils. The diameter of the head of the bayonet button is larger. The entrance hole in the pad is smaller giving a better hold for the pad on the button.

The spanner is provided for removing and replacing the barrel bush, and is fitted with two pins for removing and replacing the bayonet button.

Total length of musket (over all) . . . . .4 ft. 8 11/16 ins.
Total weight of musket (over all) . . . . .9 lb. 8 oz.
Balance from end of butt . . . . .23½ ins.

12825—Sword, staff-serjeants', pattern 1905 27 Feb 1905
   (Mark I) **L**
   Without scabbard; converted from Sword, Cavalry, pattern 1899. All dismounted services except regiments having claymores.

   **Scabbard, sword, staff-serjeants',
   pattern 1905 (Mark I)** **L**
   Steel; all dismounted services except regiments having claymores.

Patterns of the above-mentioned sword and scabbard have been sealed to govern supplies.

Sword— The sword is converted from "Swords, Cavalry, pattern 1899." The conversion consists of reducing the blade in width, thickness and length, and fitting new guard, grip plates, hand stop, grips, and pommel to the tang. A buff leather washer is placed on the blade under the guard. The guard is of steel, and is pierced with an ornamental device. The grip is 5¼ inches long.

Scabbard— The scabbard, curved to suit the blade of the sword, is of steel, fitted with a sputcheon brazed on to the mouthpiece and fixed to the scabbard by two screws. Two bands with loose rings are brazed on the scabbard, 2 3/16 inches and 10 7/16 inches respectively from the top of the mouthpiece.

The scabbard lining consists of two strips of wood held in position by the sputcheon.

See § 12825, "Sword and Scabbard, staff-serjeants', pattern 1905. (Mark I.)"

LEATHER WASHER

	feet	inches
Length of sword..................	3	2 9/16
Length of scabbard................	2	9 5/8
Length of blade from point of shoulder....	2	8 3/4
Length of sword and scabbard (with 1/3 inch buff leather washer under guard..	3	3 9/16
Balance from hilt..................	0	4 3/8
Weight of sword..................	2 lb. 2 oz.	
Weight of scabbard................	1 lb. 1 oz.	

12860—Swords—   22 Mar 1905
    Cavalry   L
    Household Cavalry   L
    Mountain Artillery   L

    Sword-bayonets—
    Pattern 1888   C
    Pattern 1903   C

Reduction in thickness of edges.
Sharpening before troops proceed on active service.

With reference to LoC 9019, 9206 and 10559— In future, only the above-mentioned swords and sword-bayonets will have their edges reduced during manufacture and repair, or be sharpened when troops proceed on active service.

12877—Cartridges, S.A., dummy, drill,   12 May 1905
    .303-inch rifles or carbines   C

Reduced diameter of holes bored through case.

With reference to LoC 12597— In future, the holes bored through the case of cartridges therein mentioned will be 1/12 inch diameter instead of 3/16 inch diameter.

Any existing dummy cartridges not yet altered in accordance with LoC 12597 will have their holes bored through the case to the new diameter as shown above. This modification will not involve any advance of numeral.

12878—Cartridges, signal, Very's—   4 May 1905
    Green   C
    Red   C
    White   C

Use extended to Land Service.

The use of the above-mentioned cartridges (LoC 5173, 7220, 8930 and 10117) having been extended to Land Service, the distinguishing letter has been altered from "N" to "C".

12990—Case, 200 scabbards, sword-bayonet,   16 Apr 1904
    pattern 1903 (Mark I)   C

Fittings not to be coated with paraffin wax.

In future manufacture of cases of the above-mentioned description (Loc 11849, 12326) the inside fittings will not be coated with paraffin wax.

12991—Case, sword-bayonet, or scabbard,   26 May 1905
    sword-bayonet pattern 1903 (Mark II)   C
    Wood; also for pattern 1888.

    Battens, plain—
    For Case, sword-bayonet, or scabbard,
    sword-bayonet pattern 1903.
    16 inches by 7/8 in. by 3/4 in. (Mark I)   C
    16 inches by 1½ in. by 3/4 in. (Mark I)   C

    Racks, sword-bayonet pattern 1903—
    No. 1 (Mark I)   C
    For 12 hilts.
    No. 2 (Mark I)   C
    For 13 hilts.

    Board, arm chest—
    16 inches by 6 1/8 in. by ¾ in. (Mark I)   C
    For Case, sword-bayonet, or scabbard,
    sword-bayonet pattern 1903.

    Patterns of the above-mentioned stores have been approved to govern future manufacture.

    The case is generally similar to the Mark I pattern (LoC 11849) but it is arranged internally to take fittings (which will be regarded as separate demandable stores as shown above) to pack "Sword-bayonets," or "Scabbards, sword-bayonet," as follows—
    1. 200 Sword-bayonets.
    2. 200 Scabbards, sword-bayonet.
    3. 100 Sword-bayonets, and 100 Scabbards, sword-bayonet.

    The "Battens," "Racks" and "Boards" detailed below will form a set for the particular packing required—

    1. 200 Sword-bayonets—   No.
       Battens, plain, 16 inches by 1½ inches by ¾ inch 4
       Racks, sword-bayonet pattern 1903—
         No. 1 8
         No. 2 8
       Boards, arm chest, 16 inches by 6 1/8 inches by ¾ inch. . .2

    2. 200 Scabbards, sword-bayonet—
       Battens, plain—
       16 inches by 7/8 inch by ¾ inch 40
       16 inches by 1½ inches by ¾ inch 4

    3. 100 Sword-bayonets and 100 Scabbards, sword-bayonet—
       Battens, plain—
       16 inches by 7/8 inch by ¾ inch 24

  16 inches by 1½ inches by ¾ inch 4
  Racks, sword-bayonet pattern 1903—
   No. 1 4
   No. 2 4
  Board, arm chest, 16 inches by 6 1/8 inches by ¾ inch ... 1

  The internal fittings for sword-bayonets only are paraffin-waxed, with the exception of "Batten, plain, 16 inches by 1½ inches by ¾ inch" which is interchangeable for packing sword-bayonets or scabbards.

Dimensions, &c.

	feet	inches
Length, over all	4	0 5/16
Width, over all	1	6 1/16
Depth, over all	1	2 3/16
Tonnage	.1791 ton.	

12992—Swivel, band—   26 Sep 1905
  Rifles, short, M.L.E.   L
  and butt.

  Swivels of the above-mentioned pattern have been approved for short rifles, for future manufacture for Land Service.

  They differ from the "Swivel, band, M.L.M. rifle, Mark II" previously used in "Rifles, short, M.L.E." in being 1/8 inch shorter, on the long side, to facilitate the insertion of the rifle in the new pattern rifle bucket (LoC 12714).

  Officers commanding Cavalry units will demand two of the new pattern swivels per short rifle on charge, and will return the old pattern swivels to store.

  Officers commanding units in possession of buff slings should, if necessary, reduce the breadth thereof regimentally to adapt them to the new swivel.

12993—Tools, adjusting foresight,   24 Nov 1904
  Rifle, short, M.L.E.—   25 Sep 1905
   Cramp (Mark I)   C
   Steel, with two adjusting screws.
   Punch, centre (Mark I)   C
   Steel.

  Patterns of the above-mentioned tools have been sealed to govern manufacture.

  The cramp, a drawing of which is attached, is used in the following manner—

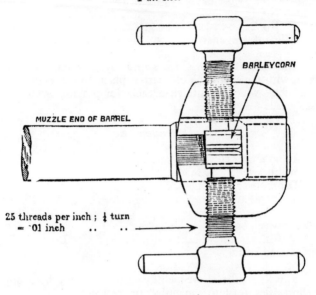

Full size.

Remove the nosecap screws and nosecap from the rifle. Turn the screws of the cramp back, and place the cramp on the barrel so that the fore-sight block is between the gap of the cramp; turn one of the screws up to the side of the fore sight it is desired to press, and move the fore sight to the desired position; remove the cramp, and replace the nosecap and screws.

When the correct position of the fore sight has been determined, the "Punch, centre" will be used to fix the foresight in position.

12994—Carrier, water-bottle, with shoulder 9 Sep 1905
   strap (Mark II)
   Brown leather. (For all services with
   Marks V or VI water-bottles).

A pattern of the above-mentioned article has been sealed to govern future manufacture.

It consists of the "Carrier, water-bottle" (LoC 11460 and 11769) with a shoulder strap somewhat similar to that in LoC 11460, but modified to suit both mounted and dismounted services.

The present pattern "Strap, shoulder, water-bottle, dismounted services," if found too long for buglers or small men, will be altered locally under the following instructions, and converted to above-mentioned pattern "Carrier, with shoulder strap"—

Instructions

The long strap to be taken off where sewn to the web; buckle to be taken out and reversed; fixed loop removed and made up to serve as runner (the loop near the buckle will not be required); a strip of leather 10 inches by ½ inch to be added; the holes from late sewing to serve as guide for length of splice.

The short strap to be cut down to 3¼ inches, and slightly skived, brass loop adjusted, and the leather chape doubled under the sewing on the top of the web and re-sewn.

Long strap to be passed through the brass loop, and buckled up with runner, then passed through the rings on the carrier and sewn on to the web in the same manner as before.

A pattern to guide conversion, and the following material will be issued on indent—

	No.
Loop, brass, 9/16 inch by 3/8 inch	1
Rivets, copper, tinned—with washers	3
Pieces, leather, brown, 10 inches by ½ inch	1

13043—Cartridge, S.A. shot— L 9 Jan 1903
   Snider, special (Mark I)

Alteration to weight of shot.

With reference to LoC 9343— The weight of the 28 lead A.A.A. shot will be 200 grains, instead of 300 grains as therein stated.

13056—Chest, Rifle, short, M.L.E. (Mark I) C 22 Mar 1905
         27 Jun 1905

Addition of double cleats.

To facilitate packing of the various patterns of sword-bayonets and scabbards in the above-mentioned chest (LoC 11808), a double cleat, in place of a single one, will in future be fitted to each side of one end of the chest.

Existing chests will be fitted with the double cleats when passing through the Ordnance Factories for repair.

13057—Pistols, Webley. (Mark III & IV) L 21 Nov 1905
   Fitted with 6-inch barrel.

It has been decided to specially fit a number of pistols as above to meet the requirements of officers and cadets desirous of purchasing such pistols from store.

13058—Lanyard, pistol, khaki.   L   13 Nov 1905
    Worsted.

A pattern of the above-mentioned lanyard has been sealed to govern future manufacture.

It differs in colour only from the previous pattern (LoC 3315) which will become obsolete when existing stocks have been used up.

13083—Cartridges, S.A. blank, .303-inch—   25 Sep 1905
    Without bullet   C
    With mock bullet   C

Nomenclature.

With reference to LoC 7519, 11317— The word "cordite" will, in future, be omitted from the designation of .303-inch S.A. blank cordite cartridges, as shown above.

13105—Cases—   10 Sep 1904
    Wood, with fittings; also for sword-   29 Aug 1905
    bayonets and scabbards.

    10 rifles, short, M.L.E. (Mark I)   C
    6 rifles, short, M.L.E. (Mark I)   C
    2 rifles, short, M.L.E. (Mark I)   C

Patterns of the above-mentioned cases have been approved to govern manufacture.

They are made of deal, with elm ends, and, with the exception of the "Case, 2 rifles" are provided with cleats and rope handles. The lids are secured by screws.

The inside fittings are of deal and beech.

Dimensions, &c.

	\\	Case.	
	10 rifles	6 rifles	2 rifles
Length over all	50.3 ins.	50.3 ins.	48.3 ins.
Width over all	21.187 ins.	14.0 ins.	6.8 ins.
Depth over all	12.0 ins.	12.0 ins.	12.0 ins.
Weight, empty	3 qrs. 4 lb.	2 qrs. 8 lb.	1 qrs. 9 lb.
Tonnage	.185 ton	.122 ton	.057 ton

13136—Carbines and rifles—   16 Feb 1906
    M.L.M. and M.L.E.
    M.M. and M.E.

Reduction in weight of "pull-off".

A reduction in the weight of the "pull-off" of the above arms, from 6 lb. minimum and 8 lb. maximum, to 5 lb. minimum and 7 lb. maximum, has been approved.

The alteration will be effected locally by armourers, in the following manner—

Instructions for reducing the weight of "pull-off", carbines and rifles, M.L.M. and M.L.E.

Test the weight of the pull-off, and if too heavy, with a piece of "Cloth, emery, fine, F," wrapped round the "File, smooth, half-round, 4-inch" slightly increase the angle of the bent of the cocking piece, as shown in dotted lines at A, Fig. 1, of the accompanying drawing, taking particular care to keep the "bent" concave and of the same radius as before; if the pull-off drags, (i.e. is not smooth) adjust the bent by smoothing with the rounded edge of the "Stone, oil, slip, Arkansas, large," Under no circumstances is the bent to be made flat or convex.

If the pull-off is still heavy, weigh the sear spring from trigger, with the bolt open (pull to move trigger 3½ lb. to 4½ lb.), and if too heavy remove the sear spring and file the nib (B, Fig. 2) flat, as shown in dotted lines at C, taking great care not to reduce the length of the nib at D; see that the sear and trigger are free on axis and replace the sear spring, and if still too heavy, remove, file the inside of the long arm at EE evenly throughout, until the required weight is obtained. Under no circumstances is the original set of the sear spring to be altered, or the weight tested from the trigger to be less than 3½ lb.

Fig. 1.

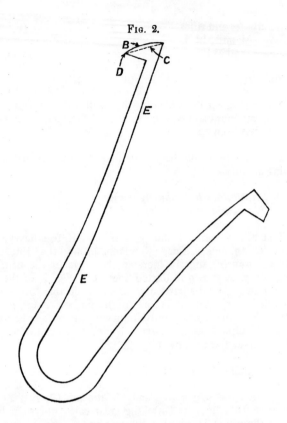

Fig. 2.

Instructions for reducing the weight of "pull-off", carbines and rifles, M.M. and M.E.

Test the weight of the pull-off (5 lb. to 7 lb.) and if too heavy remove the trigger spring and test the weight of the pull-off (2 lb. to 2½ lb.) without it, if then too heavy increase the angle of the bent of the tumbler to adjust; replace the trigger spring, and if still too heavy, remove the trigger spring and adjust by filing the underside until the necessary weight of pull-off (5 lb. to 7 lb.) is obtained.

If the pull-off without the trigger spring is too light, decrease the angle of the bent of the tumbler to pull-off between from 2 lb. to 2½ lb., replace the trigger spring, and if still too light, insert a stronger trigger spring, and if necessary file the underside and adjust to weight.

When inserting the trigger spring particular care must be taken that the spring is screwed hard and firmly down by the screw.

13137—Rifles—   23 Jan 1906
       M.L.M., Marks II & II*
       M.L.E., Marks I & I*

Riveting of the end of the piling swivel screw of rifles carried by Mounted Infantry and Imperial Yeomanry.

When rifles of the above-mentioned patterns are slung from the swivel inserted in the nosecap, the movement of the swivel is liable to work the screw loose. To prevent this, the screwed end should be lightly riveted over in the following manner—
    Turn the screw home; hold the rifle so that the head of the screw rests on the jaws of the vice; then with tang end of the "hammer, riveting, 4-oz.," lightly rivet over.

13266—Cases—   27 Feb 1906
      8 rifles, M.L.M. (Mark I)   C
      4 rifles, M.L.M. (Mark I)   C

Obsolete for future manufacture.

No more of the above-mentioned rifle cases (LoC 7030) will be made, and so soon as the existing stock is used up they will be regarded as obsolete.

Cases to hold 2, 6 and 10 rifles only, will in future be supplied.

13273—Bandolier equipment, Pattern 1903—   27 Nov 1905
                                                         26 Jan 1906
      Bandoliers—
      90 rounds (Mark II)   L
      Brown leather, with 9 pockets, triangle,
        2 buckles and runner.

      50 rounds (Mark II)   L
      Brown leather, with 5 pockets, steadying
        strap, triangle, 2 buckles and runner.

      Belt, waist (Mark I)   L
      Brown leather, with buckle and runner.

      Pocket, cartridge, 15 rounds (Mark II)   L
      Brown leather.

Patterns of the above-mentioned articles have been sealed to govern supplies.

The 90-round bandolier differs from the previous pattern (LoC 12424) in being fitted with two 1½-inch buckles and a brass triangle instead of a composite double buckle; and the 50-round bandolier from that in LoC 12389 by having a brass triangle and a second 1½-inch buckle instead of the composite buckle with loop.

The belt, waist, has been shortened by about 8 inches, and no advance in numeral is involved.

The pocket differs from the pattern in LoC 12389 in being made to open outwards; it has a strap across the top to prevent the ammunition from falling out of the pocket when open, and it is fastened by a billet from the back to a stud on the flap or cover.

13310—Cartridge, S.A., ball—   5 Mar 1906
    M.H. rifle, solid case (Mark II)   L
    Snider (Mark IX)   L

    Cartridge, S.A., blank—
    M.H. or Snider, rifle or carbine (Mark IV)   L

Distinguishing letter altered from "C" to "L".

The above-mentioned cartridges (LoC 4911, 2105 & 2332) being no longer required for Naval service, the distinguishing letters have been altered from "C" to "L" as shown above.

13313—Rifles, short, magazine Lee-Enfield.   9 May 1906

Removal of sharp edges on body.

It has been found that the sharp edges of the charger guide on the left side of the body, and the charger guide stop on the right side of the body, injure the clothing and bandolier, when the rifle is carried at the slope. To prevent this, the following alteration will be carried out—

> The sharp edges of the charger guide and stop will be filed off, as shown at A and B of the accompanying drawing, great care being taken not to reduce the bearing surface on the stop for the charger guide of the bolt-head, or remove the browning on

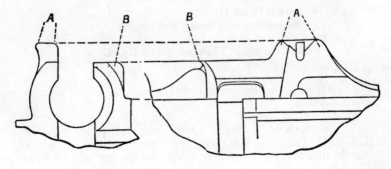

the body, more than necessary. After the corners and sharp edges have been rounded, the bright surfaces will be painted with a mixture used for temporary blacking, according to the instructions given on page 52, "Instructions for Armourers, 1904."

Rifles in the hands of troops will be altered regimentally, and those in store locally.

Naval Service rifles in Gunnery Schools and at Marine Divisions will be altered by the armourers belonging to these establishments.

11314—Slide, sight, wind-gauge, M.E., M.L.M., or   16 May 1906
     M.L.E. rifles (Mark I)
     Consisting of elevation slide and spring,
     wind gauge and spring.

A pattern of the above-mentioned slide has been sealed, to govern supplies on repayment, for the use, in voluntary practices, of officers, non-commissioned officers and men.

Demands for such numbers of these slides as may be required should be put forward in the usual way by officers commanding.

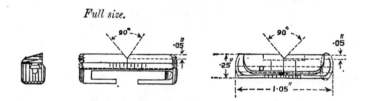

Full size.

The slide consists of an elevation slide, fitted with a gib spring at the side to tension the slide on the leaf; and a wind gauge, which is dovetailed in the elevation slide, and is capable of lateral adjustment for ¼ inch to the right or left. A transverse spring is fitted within the dovetail; this spring limits the movement and regulates the tension of the wind-gauge slide.

The wind gauge is provided with two V notches, one for use at ranges from 200 to 400 yards with the leaf down, and the other for use at ranges from 500 to 1,800 yards with the leaf raised. It is also marked with the centre lines under each V notch, which act as pointers to the wind-gauge scale on the back edge of the elevation slide. This scale has divisions of .0388 inch on the back edge, marked by alternate long and short lines, each division representing a wind allowance of 6 inches for each 100 yards the firer is distant from the target. To fit this slide the ordinary cap screw and slide must be removed from the leaf, and the wind-gauge slide put on in their place.

To remove the wind gauge from the elevation slide, move the wind gauge to the extreme right, then with a piece of wire, of a suitable size inserted in the hole in the top of the wind gauge, depress the wind-gauge spring, at the same time pressing the wind gauge to the right.

The pattern introduced by LoC 6762 is declared obsolete.

13465–Charger, .303-inch cartridges (Mark II) C 24 Apr 1906
Steel; to hold 5.

A pattern (drawing C.I.W./1073) of the above-mentioned charger has been sealed to govern future manufacture.

It differs from the Mark I pattern (LoC 11753) in having a spring stop formed at each end, and in being strengthened by having three ridges on the base. The numeral (II) is shown on the side of the charger.

The Mark II charger involves no change in the packing.

13470–Bottle, oil (Mark IV) C 22 Mar 1906
Brass, with washer and stopper; all .303-
inch magazine carbines and rifles, including
Rifles, short, M.L.E. that have traps to butt
plates.

A pattern of the above-mentioned oil bottle has been sealed to govern future manufacture.

It differs from the Mark III oil bottle (LoC 10122) in having a flat instead of convex bottom.

13509–Rifle, short, magazine, Lee-Enfield, C 26 Sep 1905
 (Mark I) 2 Jul 1906

Rifle, short, magazine, Lee-Enfield, C
converted. (Mark II)
Converted from Rifles, magazine, Lee-Enfield,
Marks I and I*, and magazine, Lee-Metford,
Marks II and II*.

With reference to LoC 11947, 11949 and 12992, and the rifles &c. therein described— Patterns of the above-mentioned rifles have been sealed, embodying all alterations and additions which have been approved.

Barleycorns, foresight –   24 Feb 1905
                                                            2 Mar 1905

The top edge of the wind gauge is hardened to prevent injury. The rear end of the sight leaf is bored to receive two spiral springs to tension the fine adjustment, and a new "screw, fine adjustment", is fitted, having a larger head and finer milling, with a tension spring fitted under the head to prevent the screw jarring loose on firing. When the present stock of "screws, fine adjustment" is used up they will be replaced by the new pattern screw with spring.

Guards, hand –   14 Feb 1905

The ends of the two handguards where they meet are vertical, and the rear handguard is fitted with a double spring, fixed by two longer rivets placed in a central position in the length of the spring. Handguards with single springs will be exchanged for those with double springs. Instructions for altering the fore-ends locally to take these handguards are given on page 11 of Supplement to Armourers' Instructions, 1904, for the care, preservation and repair of the short rifle.

Guard –   30 Sep 1905
                                                                  17 Apr 1906

The guard is provided with a lug at the front to carry the sling swivel and screw, and the magazine link loop has been removed. Trigger guards with the new swivel attachment will be issued and fitted locally in accordance with instructions issued with circular letter to General Officers Commanding. The trigger guards and magazine links removed from the rifles will be returned to store.

Link, magazine –   30 Sep 1905

The magazine link is omitted from these rifles.

Magazine –
The magazine link loop has been removed from the magazine.

Plate, keeper, stock bolt –
Stock-bolt keeper plates with the sides of the locking groove raised 1/8-inch have been fitted. All spare stock-bolt keeper plates will in future be issued to this pattern when a new one is required. Armourers will alter the recess at the rear end of the fore-end to receive it according to instructions given on page 9 of Supplement to Armourers' Instructions, 1904.

Screw, nut, keeper striker, and nut –   3 Nov 1904

A stronger "screw, nut, keeper striker" having a suitable nut, has been fitted to the cocking piece. When the present stock of "screws and nuts, keeper striker" is used up they will be replaced by the new pattern screw and nut.

Spring, platform, magazine –   25 Jan 1905

A retaining nib has been raised on the top bend of the spring to prevent any movement of the platform on the spring. The magazine platform spring of rifles in store and in the hands of troops will be altered by fitting a nib, which will be riveted on the top bend. Magazine platforms and springs with nib fitted will be issued from store and fitted to magazines locally, the platforms and springs removed being returned to store.

Stock, fore-end—   10 Apr 1905
        24 Oct 1905

The "stock, fore-end" has been recessed to receive a stud and spring to centre the barrel in the nosecap. It has also been suitably recessed for the stock-bolt keeper plate with higher locking sides, and for the double spring of hand-guards.

Fore-ends fitted with the new stud and spring are marked with the letter "S" below the nosecap.

A screwed pin for fore-ends is fitted, replacing the rivet fore-end with two washers in "Rifle, short, M.L.E. Mark I."

Swivels, band, outer; guard and butt—   14 Apr 1905

Swivels 1/8-inch shorter on the long side (referred to in LoC 12992) have been fitted for Land Service rifles only.

In rifles for Naval Service the long swivels will be used.

Swivel, piling—   30 Sep 1905

A piling swivel is fitted in the nosecap of all rifles, except for Cavalry.

In addition to the components shown in the above-quoted paragraphs, List of Changes, as special to "Rifles, short, M.L.E.," the following, also, are special, consequent on change of pattern of components, and the addition of new components—
    Rivets, spring handguard (2)
    Springs, fine adjustment (2) †
    Spring, guard, hand, rear, double
    Spring, screw, fine adjustment †
    Spring, stud, fore-end †
    Stud, fore-end †
    Swivels, band, outer— guard and butt (2) +

Components marked thus (†) are additional.
Components marked thus (+) are special for Land Service only.

The rifles in store at Weedon will be altered as regards the fitting of "springs, fine adjustment," "screw, and spring, screw, fine adjustment," "spring, platform, magazine," and rear handguard with double spring. They will also be fitted with the trigger guard with swivel attachment, and have the magazine link and link loop on magazine removed.

**13510—Rifle, short, magazine, Lee-Enfield,   2 Jul 1906
converted (Mark I)   L**
Converted from Rifle, magazine,
Lee-Metford, Mark I*.

Obsolete.

With reference to LoC 11948 and 12091, and the rifle above-mentioned therein described— No rifles of this pattern have been issued or will be manufactured.

13511—Screw, keeper, striker, M.L.E. rifles, short   C   17 Aug 1906

A keeper screw for retaining the striker in cocking pieces of "Rifles, short, M.L.E.," which can be turned by a coin, has been approved to replace the "nut, screw and spring, keeper, striker," which will become obsolete as the stock is used up.

The striker keeper screw will be fitted locally, and the following tools will be supplied from store to enable the armourers to do the work—
For Land Service only—
Broaches   2   One in handle and one spare.
Taps   3   One in handle and one spare.
Tool, recessing 1   In handle.

For Naval Service— The tools will only be supplied to Training Establishments on demand.

Instructions for fitting "screw, keeper, striker."

Remove the cocking piece from the bolt, and, with the broach, open the small hole at the bottom of the nut recess until the pin in the end of the recessing tool fits. Tap the hole (taking care to keep the tap in line with the cocking piece) then with the recessing tool remove the burr raised by the broach and tap at the bottom of the recess, taking care not to deepen the recess, insert the "screw, keeper striker," and screw home in recess and examine to see that it fits properly; remove the screw and assemble bolt.

**13544—Cartridges, S.A., blank, .303-inch—   25 Sep 1906
    Without bullet (Mark VI)   N**
Solid case.
With mock bullet (Mark VI)   L
Solid case.

Removal of mock bullet from cartridges in Naval Service. Nomenclature.

With reference to LoC 11317— It has been decided to remove the mock bullet from all Mark VI .303-inch blank cartridges in Naval Service.

This service will be carried out on board H.M. ships in accordance with instructions contained in Admiralty Circular Letter, G 9263/06, dated 16th July, 1906, which has been issued to all concerned.

The bullets will be removed from Naval cartridges in store at Woolwich before issue, and the cartridges will then be packed and issued in metal-lined cases as follows—

Package	No. of cartridges without mock bullets.
¼ M.L. case	1,450
½ M.L. case	3,400

In consequence of the foregoing decision the nomenclature and distinguishing letter of Mark VI .303-inch blank cartridges have been arranged as now shown.

13549—Rifles, short, magazine, Lee-Enfield   12 May 1906
27 Sep 1906

Removal of sharp edge on locking bolt.

It has been found that the sharp edge of the locking bolt injures the clothing and bandolier when the rifle is carried at the slope. To prevent this the following alteration will be carried out—

> Remove the locking bolt from the rifle and file off the sharp edge as shown at A on the accompanying drawing. After the sharp edge has been rounded, the bright surface will be painted with the mixture used for temporary blacking, according to instructions given on page 52, "Instructions for Armourers, 1904."

Full size.

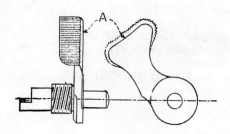

Rifles in the hands of troops will be altered regimentally and those in store, locally.

Naval Service rifles in Gunnery Schools and at Marine Divisions will be altered by the armourers belonging to these establishments.

13577 – Rifle, short, magazine, Lee-Enfield (Mark I*) C 27 Mar 1906
2 Jul 1906
9 Jul 1906
6 Jul 1906
17 Aug 1906
12 May 1906

A pattern of the above-mentioned rifle has been approved to govern present manufacture and alteration of Mark I rifles as may be ordered.

It differs from the "Rifle, short, M.L.E., Mark I" (LoC 13509) in the following particulars—

The butt plate is made of gunmetal, and is fitted with a trap for insertion of the oil bottle and pullthrough in the butt.

The stock, butt, is recessed to receive a small bracket fixed by two screws, to carry the sling swivel, and also for the gunmetal butt plate. The division between the two upper lightening holes is cut away to allow of the insertion of the pullthrough weight. A leather wad is inserted on top of the stock bolt to act as a buffer for the oil bottle.

A magazine case (No. 2) which is deeper at the front to facilitate loading, is fitted, and an auxiliary spring (No. 2) having a straight instead of curved end, is fitted to the magazine case. The magazine case and auxiliary spring will be marked with the figure "2" for purposes of identification.

To prevent the swivel screws working loose by the movement of the swivels, the swivel screws are bored longitudinally at the screwed end to enable the end of the screw, when screwed home, to be expanded by the armourer with the centre punch issued as part of the foresight adjusting tools.

A keeper screw for striker, which can be turned by a coin, is fitted to the cocking piece. This screw replaces the nut, screw and spring, keeper striker.

The sharp corners of various components have been rounded to prevent injury to equipment and clothing.

The following is a list of new and additional components—

Bracket, swivel, butt	Common to "Rifles, short, M.L.E. Marks I* and converted II*.
Screws, bracket, swivel, butt (2)	Same as "Screw, strap, butt plate, iron."
Butt plate	Special.
Butt plate trap	Special.
Spring, trap, butt plate	Special.

Screw, spring, trap, butt plate	Special.
Pin, axis, trap, butt plate	Special.
Magazine case, No. 2 †	Interchangeable in all short rifles.
Spring, auxiliary, platform, magazine, No. 2 †	For magazine case, No. 2.
Wad, stock bolt	Same as "Wad, stock bolt, M.L.M."
Stock butt	Special.
Screw, swivel †	Interchangeable in all short rifles.
Screw, keeper, striker †	Common to "Rifles. short, M.L.E. Marks I* and converted II*. Will be fitted to all short rifles, locally, as stock of nuts, screws, and springs are used up.

† Rifles, short, M.L.E., Marks I* and converted II* of early manufacture were not fitted with these components.

13578—Rifle, short, magazine Lee-Enfield,   2 Jul 1906
      converted (Mark II*)   **C**
      Converted from Rifles, magazine Lee-Enfield, Marks I and I*, and magazine Lee-Metford, Marks II and II*.

A pattern of the above-mentioned rifle has been approved to govern future manufacture.

It differs from the "Rifle, short, M.L.E., converted, Mark II" (LoC 13509) in the following particulars—

The stock, butt, is recessed to receive a small bracket fixed by two screws, to carry the sling swivel. The stock bolt hole is of a depth suitable for "stock bolt No. 2, M.L.E. rifle" which will be used. The division between the two upper lightening holes is cut away to allow of the insertion of the pullthrough weight.

The "butt plate, M.L.M. rifle, Mark II" having a strap at heel for regimental marks, and a trap for the insertion of the oil bottle and pullthrough in the butt, is retained.

The "wad, stock bolt, M.L.M." is retained.
The "screw, strap, butt plate, iron" is retained.
The "marking disc and screw" are omitted.

A magazine case, No. 2, which is deeper at front to facilitate loading; is fitted, and an auxiliary spring (No. 2), having a straight instead of curved end, is fitted to the magazine case. The magazine case and auxiliary spring will be marked with the figure "2" for purposes of identification.

To prevent the swivel screws working loose by the movement of the swivels, the swivel screws are bored longitudinally at the screwed end to enable the end of the screw, when screwed home, to be expanded by the armourer with the centre punch issued as part of the foresight adjusting tools.

A keeper screw for striker, which can be turned by a coin, is fitted to the cocking piece. This screw replaces the nut, screw and spring, keeper striker.

The sharp corners of various components have been rounded to prevent injury to equipment and clothing.

See footnote (†) (LoC 13577) for "Rifle, short, M.L.E., Mark I*" as regards certain components not being fitted to converted Mark II* rifles.

13579–Tool, removing striker, Rifle, short, M.L.E. C 28 Sep 1906

A pattern of the above-mentioned tool has been sealed to govern manufacture and supply.

The tool is for use of armourers when stripping bolts of "Rifles, short, M.L.E." the strikers of which are very tight in the cocking-piece, as in such cases removal of the striker by means of the bolt-head is liable to injure the striker lug recess and thread on the tenon of the bolt-head.

The tool has a ¼-inch hole drilled through and a slot cut on top of the head to enable a tommy or screwdriver to be used for turning it.

13611–Carbines & Rifles, M.L.M. & M.L.E. 3 Jan 1907

Fitting of spare bolt-heads.

Spare bolt-heads for the above-mentioned arms will, in future, be left .005 inch longer in front, and will be fitted in accordance with instructions given in LoC 12611.

The present stock of spare bolt-heads will, however, be used up and officers commanding will specify on demands how many of those demanded are required to be longer at front for rifles in which a bolt-head of ordinary length will not prevent the bolt turning over the .072-inch gauge for "distance of bolt from end of chanber."

13612–Cut-off– 25 Oct 1906
Rifle, short, M.L.E. (Mark I) C

Use extended to Land Service.

The use of the above-mentioned cut-off (LoC 12129) has been extended to Land Service, and the distinguishing letter altered from "N" to "C".

The cut-off will be fitted to arms in the hands of troops, and in store, by regimental and Army Ordnance Department armourers, respectively, in accordance with the following instructions—

Remove the fore-end, cut-off screw, &c.

Assemble cut-off to body. (Should the cut-off be found to bear, between the stops at the wide end, against the end of the cut-off slot in the body, remove sufficient of the metal, with a smooth file, to clear the body.)

Test for correct action. (Should the cut-off be difficult to open, remove the cut-off, and with a small flat scraper slightly round the sharp upper inside edge of the cut-off slot in body.)

Remove magazine, screws, trigger guard, fore-end, &c., and lubricate with "composition, preserving arms," the body, barrel, and nosecap, and the barrel groove in fore-end and handguards, and then re-assemble the arm.

Demands for cut-offs will be put forward in accordance with instructions which will be communicated to Chief Ordnance Officers by C.O.O., Weedon.

13642–Cartridge, aiming tube, R.F. (Mark I)   L   29 Nov 1904
  Small arms   6 Jul 1905

1. New pattern.

Cartridge, aiming tube, C.F. (Mark II)   L
Morris, small arms.

2. Nomenclature.

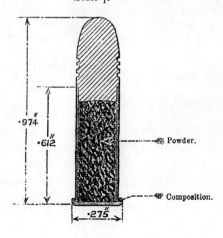

Cartridge, aiming tube, R.F. (Mark I.).
Scale 2/1.

1. A drawing (R.L. 13881) of the above-mentioned R.F. (rim fire) cartridge has been sealed to govern manufacture.

The case is made of solid-drawn copper-zinc alloy, the fold in the rim being charged with composition as shown in the accompanying drawing.

The lead bullet weighs 39.9 grains, and has three roughened cannelures round it to retain the lubricant. The charge is 4.7 grains of black powder of ordinary R.F.G. 2 proportions.

The cartridges will be packed 100 in a cardboard box, 10 cardboard boxes in a tin, and 10 tins in a wood box = 10,000 rounds in all. The latter box is made of ¾-inch deal, the top and bottom secured by eight and fourteen brass screws respectively, and the ends are fitted with wood cleats amd rope handles.

Dimensions of box.

Exterior for stowage— length . . . . .17.55 inches.
width . . . . .13.8 inches.
depth . . . . 6.9 inches.

2. Consequent on the introduction of the "Cartridge, aiming tube, R.F., Mark I", the nomenclature of the "Cartridge, aiming tube" (LoC 6391) has been amended by the addition of "C.F." (central fire), as shown above.

13649 – Rifles, short, M.L.E., A.T. (Mark I) L   13 Nov 1906
For .22-inch rimfire cartridge aiming tube,   1 Feb 1907
and for drill purposes.

Rifles, short, M.L.E., $\frac{A.T.}{N.I.}$ (Mark I)   L

For .22-inch rimfire cartridge aiming tube; non-interchangeable; part of 1,000 experimental "Rifles, short, M.L.E.," that were issued for trial.

1. In future, in the Land Service, only one class of "Rifles, Short, M.L.E." will be issued for aiming practice and drill purposes. These rifles will be fitted with boltheads and strikers suitable for .22-inch rimfire aiming-tube cartridges; the striker hole in the face of the bolthead being bored .10 inch eccentric to the axis of the bolt, the point of the striker being arranged suitably, and the rear end of the extractor being altered to clear the special striker.

The barrels of these rifles will be so chambered that Service ball or blank ammunition cannot be fired in them. These rifles will be marked A.T. (aiming tube) on the body, barrel, bolt, and stocks, butt and fore-end, and classified as such. The components of these rifles (with the exception of the barrel, bolthead, striker and extractor) will be interchangeable according to the "Mark" of rifle from which they are prepared.

2. Certain of the rifles are not suitable for instruction in the parts of the rifle, as the components (particularly the sights) differ from the service pattern. Two patterns of windgauge-backsights are fitted to these rifles, and are placed in two positions.

With the exception of a few small components, such as screws, &c., and the components mentioned below, the components of these rifles are non-interchangeable, and will not be replaced as they become unserviceable. The rifles will, consequently, not be repaired (except for such repairs as can be effected locally without the supply of new components), except in the case of the following, which are interchangeable—
 Heads, breech bolt, M.L.M.R., Mark I*, R.F.
 Strikers, M.L.M.R., Mark I*, R.F.
 Extractors, M.L.E.R.S., Mark I, R.F.

These rifles will be marked A.T. /N.I. (Aiming tube / Non-interchangeable) on the body, barrel, bolt, and stocks, butt and fore-end, and classified as such. The boltheads, strikers and extractors of the above rifles (A.T. and A.T./N.I.) are interchangeable.

N.B. The strikers of these rifles should not be stripped by means of the bolthead.

13650—Tube, aiming, .22-inch, M.L.M. rifle,   29 Nov 1904
 converted   L
 Also Rifles, M.L.E.; for R.F. cartridges;
 converted from Tube, aiming, M.L.M. rifle,
 Morris; including extractor and screw, set
 nut and washers.

 Tube, aiming, .22-inch, M.L.M. rifle
  (Mark I)   L
 Also Rifles, M.L.E.; for R.F. cartridges; including extractor and screw, set nut and
  washers.

 Tube, aiming, .22-inch, M.L.E. rifle,
  short (Mark I)   L
 For R.F. cartridges; including extractor
  and screw, set nut and washers.

 Bolt, breech, M.L.M. rifle, Mark I*; for
  R.F. aiming tube cartridges   L
 Assembled; converted from Bolt, breech,
  M.L.M. rifle, Mark I*.

**Bolt, breech, M.L.M. rifle, Mark II; for
R.F. aiming tube cartridges   L**
Assembled; converted from Bolt, breech,
M.L.M. rifle, Mark II.

**Bolt, breech, M.L.M. rifle, Mark II*; for
R.F. aiming tube cartridges   L**
Also Rifles, M.L.E.; assembled; converted
from Bolt, breech, M.L.M. rifle, Mark II*

**Bolt, breech, M.L.E. rifle, short; for
R.F. aiming tube cartridges   L**
Assembled; converted from Bolt, breech,
M.L.E. rifle, short.

Patterns of the above-mentioned aiming tubes and breech bolts have been sealed to govern manufacture and conversion.

Tubes

The new tubes are similar in size to the tubes described in LoC 6602 and 6723, but differ in the size of the chamber and extractor which are arranged for the .22-inch rimfire cartridge.

Length over all of "Tubes, aiming, .22-inch
M.L.M. rifle, converted, Mark I" . . . . 30.94 inches
Length over all of "Tube, aiming, .22-inch,
M.L.E., rifle, short, Mark I" . . . . 26.0 inches

The instructions for use given in LoC 6602 will apply to these tubes, with the exception of that part of the last paragraph referring to the cleaning rod. The cleaning rod and brush, and packing washers described in LoC 7880 and 9178 are suitable for these tubes. These tubes can only be used in rifles fitted with a special bolt adapted for rimfire ammunition. The removeable components of these tubes are interchangeable.

Bolts, breech

The "Bolt, breech, rifle, M.L.M., Mark I*," for .22-inch R.F. aiming-tube cartridge is converted from "Bolt, breech, rifle, M.L.M., Mark I*". The conversion consists in removing the cover stops and a portion of the rib, plugging the bolt cover and bolt-head screw holes, and fitting and brazing in a screwed bush to receive the new bolt-head, which is arranged to receive a new striker, the point of which is placed .10-inch eccentric to the axis, suitable for .22-inch rimfire cartridges. The extractor is converted to clear the new striker.

The "Bolts, breech, rifles, M.L.M., Marks II and II*" and "Rifle, short, M.L.E.," for .22-inch R.F. aiming-tube cartridges are converted from "Bolts, breech, rifles, M.L.M., Marks II and II*", and "Rifle, short, M.L.E." respectively. The conversion consists in altering the extractors to clear the new strikers and fitting new bolt-heads and strikers suitable for .22-inch rimfire cartridges.

These bolts should not be stripped by means of the bolt-head. The bolts will be marked "A.T." (aiming tube) on the top of the lever.

The following table shows the interchangeability of the components of the breech bolts—

Bolt, breech, M.L.M. rifle—
Mark I* R.F. — Special.
Mark II R.F. — Special.
Mark II* R.F. — Interchangeable in rifles, M.L.M., Mark II* and M.L.E. Marks I and I*

Bolt, breech, M.L.E. rifle, short, Mark I R.F. — Interchangeable in all short rifles.

Heads, breech bolt, M.L.M. rifle, Mark I* R.F. — Interchangeable in all R.F. bolts, breech.

Extractor— M.L.M. rifle—
Mark I* R.F. — Special.
Mark II R.F. — Interchangeable in rifles, M.L.M., Marks II and II*, amd M.L.E., Marks I and I*

Extractor— M.L.E. rifle, short—
Mark I R.F. — Interchangeable in all short rifles.

Strikers, M.L.M. rifle—
Mark I* R.F. — Interchangeable in rifles, M.L.M., Marks I* and II, and rifles, short, M.L.E.
Mark II* R.F. — Interchangeable in rifles, M.L.M., Marks II*, and M.L.E. Marks I and I*.

Springs—
Main, M.L.M. — Interchangeable in all.
Extractor, M.L.M. — Interchangeable in all.

Screws, extractor, M.L.M. — Interchangeable in all.

Cocking-piece— M.L.M. rifle—
Mark I* — Special.

Mark II	Special.
Mark II*	Interchangeable in rifles, M.L.M., Mark II* and M.L.E., Marks I and I*.
Cocking-piece— M.L.E. rifle, short, Mark I	Interchangeable in all short rifles.
Screw, keeper, striker, M.L.M. rifle—	
Mark I*	Interchangeable in rifles, M.L.M., Marks I* and II.
Mark II*	Interchangeable in rifles, M.L.M., Mark II* and M.L.E., Marks I and I*.
Screw, nut, keeper, striker, M.L.E. rifle, short, Mark I	Interchangeable in all short rifles.
Nut, keeper, striker, M.L.E. rifle, short, Mark I	Interchangeable in all short rifles.
Spring, nut, keeper, striker, M.L.E. rifle, short, Mark I	Interchangeable in all short rifles.
Cover, breech bolt, M.L.M. rifle, Mark II	Interchangeable in rifles, M.L.M., Marks II and II* and M.L.E., Marks I and I*.
Catch, safety, M.L.M. rifle, Mark II*	Interchangeable in rifles, M.L.M., Mark II, and M.L.E. Marks I and I*.
Spring, pin, catch, safety, M.L.M. rifle, Mark II*	Interchangeable in rifles, M.L.M., Mark II, and M.L.E. Marks I and I*.
Pin, catch, safety, M.L.M. rifle, Mark II*	Interchangeable in rifles, M.L.M., Mark II, and M.L.E. Marks I and I*.

13651—Tube, aiming, .22-inch, M.L.E. rifle, short (Mark I*)   13 Nov 1906   L

For rimfire cartridges; including extractor and screw, set nut and washers.

A pattern of the above-mentioned tube has been sealed to govern future manufacture.

It differs from the Mark I tube described in LoC 13650 in the following particulars—

The cartridge-head recess of the extractor is continued towards the centre, so that it grips nearly one-half the circumference of the cartridge head, and thus improves the extraction of the empty cartridge case.

The rim of the extractor has also been extended to the left, so that when the right side of the rim of the extractor becomes thin by wear, the worn portion of the rim can be filed away by the armourer, and the unworn portion brought into position when the tube is adjusted in the rifle.

13679—Bag, armourers (Mark I) L   16 Feb 1907
   Leather, with shoulder strap and buckle,
      and two buckles for flap straps; for tools
      and components for repair of small arms
      in the field.

   Box, rod, tool, clearing (Mark I) L
   Wood, with lid and six wood screws; for
      repair of small arms in the field.

Patterns of the above-mentioned articles have been approved to govern manufacture.

The bag and box, which are to be stored with mobilization equipment, are for use of armourers of units of Cavalry, Infantry, and Mounted Infantry, for which mobilization equipment is stored, for carrying such tools and components as are mostly required for repair of small arms in the field.

Bag— The bag, which is of leather, with shoulder straps, &c., fitted, is arranged to carry the following tools and components, the components being carried in the small inside pocket—

Tools.
Brace, armourers, Mark III.
Brace, armourers—
   Bit, screwdriver, stock-bolt, M.L.M.
Driver, screw, extractor, axis.
Driver, screw, large.
Driver, screw, small.
File, smooth, flat, 6-inch (with handle).
Hammer, riveting, 4-oz.
Implement, action, R.S.M.L.E.
Pincers, armourers.
Tool, removing striker, R.S.M.L.E.
Tools, clearing, .303-inch arms—
   Bit, screw.
   Bit, spoon.
   Bush, bit, screw.

Tool, removing wad, stockbolt.
Cloth, emery, F. (2 sheets).

Components.

5 cocking pieces.
1 head, breech bolt.
10 screws, swivel.
5 sears.

Box, rod, tool, clearing— The box, which is of wood with a lid fixed by six screws, is provided for the "Rod, No. 2— Tools, clearing, .303-inch arms."

(On mobilization, the tools and components will, in the case of units of Cavalry and Infantry, be taken from those already in possession of the units, but in the case of units of Mounted Infantry they will be stored as mobilization equipment).

13680—Carbines—   18 Jan 1907
    Martini-Enfield—
        Artillery   L
        Cavalry   L
    Martini-Metford—
        Artillery   L
        Cavalry   L
    Except Marks I* and II*,

Removal of protecting wings on foresight.

Approval has been given for the removal, locally, by regimental and Army Ordnance Department armourers, of the protecting wings on the foresights of M.M. and M.E. carbines, excepting M.M. Cavalry carbines, Marks I* and II*.

The removal will be carried out according to the instructions given in LoC 11713, the block of the foresight being filed to a suitable width to fit the slot cut below the hood on the left side of the steel sight protectors. The bright surfaces will be painted with the mixture used for temporary blacking, according to instructions given on page 52 of "Instructions for Armourers, 1904."

"Carbines, Cavalry, M.M., Marks I* and II*," in possession of units, or held by Army Ordnance Department, should be exchanged for "Carbines, Artillery, M.E.," with wings of the foresight removed, a demand being put forward on the Chief Ordnance Officer, Weedon for the number required.

No further issues to troops of "Carbines, Cavalry, M.M., Marks I* and II*" will be made.

13713—Valise, equipment, pattern 1888, buff—   23 Apr 1907
Pouch, ammunition, .303-inch, 50 rounds,
pattern 1894, Mark I.
Opening outwards.

Modification.

1. A sample of the above-mentioned pouch has been modified to suit charger packed ammunition (LoC 11753), and pouches so altered will be described as—

Valise, equipment, pattern 1888, buff—
Pouch, ammunition, .303-inch, 20 rounds.
Opening outwards.

2. When specially ordered, pouches of this description in possession of units will be altered in accordance with the following instructions—

Instructions.

Cut the stitching which secures the tubing, and remove it. Insert two pieces of brown leather, and secure each piece with three rivets, one at each end and one in the centre, the distance between the rivets being 2¾ inches, thus forming loops for four chargers.

The top edge of the brown leather riveted to the front part of the pouch to be 3¾ inches from the top edge of the flap, and that riveted to the back to be 1 inch from the top edge of the pouch where the brass loop is secured.

A sample pouch to guide alteration will be issued to each unit on indent, together with the following stores—

Stores per pouch to be altered.

                                                                  No.

Pieces, leather, brown, 6½ inches by ¾ inch 2
Rivets, copper, tinned, with washers, 3/8 inch, gauge No. 11 6

13715—Bucket, rifle (Mark IV)   3 Apr 1907

LoC 11208 and 11856. Shortening of "Straps, suspending".

The suspending straps of the above-mentioned bucket having been found too long for the convenient carriage of the rifle, the sealed pattern has been altered as follows—

The buckle pieces have been shortened 2 inches, and the straps 2 inches, leaving the latter 12 inches in length, the loops being removed and sewn lower down (see drawing).

This will guide future manufacture and the alteration of exist-

See § 13715, "Bucket, rifle. (Mark IV.)

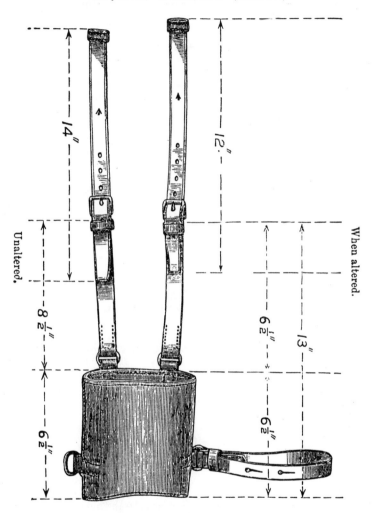

ing store of this pattern. "Buckets, rifle," in possession of units will be altered regimentally as follows—

Instructions.

Mark IV, the body of which is 6½ inches deep— as above. The straps must not be shortened by cutting the point.
Marks II and III, the bodies of which are 8½ inches deep— to have the buckle pieces shortened 4 inches, and the straps as above.

After alteration, each pattern measured from the bottom of the bucket to the top of the buckle piece (excluding buckle) should be 13 inches.

The undermentioned material will be issued on demand, viz.—
Thread, flax, fine. . . . . 1 oz. per 25 rifle buckets.
Wax, black (H. & S.) . . . . ½ oz. per 25 rifle buckets.

Rifle buckets in Ordnance and Mobililization Stores will be altered locally.

13737—Cartridge, S.A., blank, .303-inch,   22 Feb 1907
    with mock bullet, Mark VI   L
    Solid case.

Restricted use.

    Cartridge, S.A., blank, .303-inch,
        without bullet, Mark V   C
    Solid case.

Extended use.

    Cartridge, S.A., blank, .303-inch,
        with mock bullet removed, Mark VI.   C
    Solid case.

With reference to W.O. circular letters, dated 30th June, 1906, and 15th February, 1907, under authority of 104/Gen. No./3136— It has been decided to restrict the use of .303-inch blank with mock bullet, Mark VI (LoC 11317, 13544) in the Land Service, and to remove the mock bullets of surplus cartridges.

Mark V cartridges (LoC 7519) will be demanded by commands where no surplus with Mark VI with mock bullet exists.

The instructions for removing the bullets are as follows—
Place the mock bullet two-thirds of its length in a hole, slightly larger in diameter than the bullet and drilled in a hard wood bench, and force the cartridge to one side; the bullet will fall through the hole, leaving the cartridge in the hand.
Or—
Hold the cartridge in the left hand, and with a pair of suitable pliers in the right hand, remove the mock bullet by forcing it first to one side and then to the other.

Wrappers and slips for rebundling the converted cartridges can be obtained on demand. The converted cartridges pack similarly to the Mark V cartridges (LoC 7519).

13738—Cartridge, S.A., blank, .303-inch,   23 Apr 1907
    with mock bullet removed, Mark VI   C
    Solid case.

With reference to LoC 13544— The nomenclature of the converted cartridge has been altered to read as above.

13742—Guide, charger head, breech bolt, M.L.E.   9 Feb 1907
    rifle, short, Mark I, No. 2   C

    Screw, stop, charger guide, M.L.E. rifle,
    short, Mark I, No. 2   C

    Cut-off, M.L.E. rifle, short, Mark I*   C

    Plate, butt, M.L.E. rifle, short,
    Mark I*, No. 2   C

All M.L.E. rifles, short.

Patterns of the above-mentioned components have been approved to govern future manufacture.

Charger guide, No. 2.

The charger guide differs from the existing pattern in being fitted with a larger and stronger screw, and is only interchangeable with the existing pattern when fitted with the No. 2 screw.

Charger guide stop screw, No. 2

The screw differs from the existing pattern by being larger in diameter and longer in the shank, a flat being on the shank to clear the slide on the bolt-head. The screws, as spare parts, will be issued soft and without the flat on the shank. They will be fitted by armourers in the following manner—

    After the "charger guide, No. 2" is fitted, hardened, and adjusted according to instructions givern in LoC 12611, it will be removed from the bolt-head, and the No. 2 screw screwed home, the screwdriver slot being brought to the vertical position (this to prevent the head of the screw being broken through the slot by the jar when closing the breech); the shank of the screw will then be marked by a scriber showing the position for the flat, the screw will be removed and the flat filed on the shank sufficient to clear the slide on the bolt-head, the screw will then be case-hardened and the charger guide and screw assembled to the bolt-head.

This pattern of screw will not interchange with the existing pattern.

When the stock of the existing pattern charger guide and stop screw is used up, serviceable charger guides, the stop screws of which have become unserviceable, will be returned to store for conversion, and will be exchanged for No. 2 charger guides and stop screws.

Cut-off.

The cut-off differs from the existing pattern (LoC 12129) in being thicker and stronger at the joint end. The instructions given in LoC 13612 will be followed when fitting this cut-off to rifles in the hands of troops and in store.

Butt plate, No. 2.

The butt plate differs from the existing pattern in the following particulars—

The axis pin is fitted in a similar manner to that of the "M.L.M. rifle, Mark II."

The butt plate assembled is interchangeable with the existing pattern, but the trap and axis pin will not interchange with those of the existing pattern.

For Naval Services—

Instructions will be issued by the Admiralty when these alterations are to be made to existing rifles at Naval Ordnance Depots.

13754—Belts, waist—   20 Jun 1907
    Black, warrant officers   L
    and S.S. rifles, japanned, with snake hook.

    Buff, warrant officers—   L
    Artillery
    G. S.
    and S.S., dismounted services; also Engineers, with plate and universal locket.

Part of LoC 12223 cancelled.

That portion of LoC 12223 directing the above-mentioned articles to become obsolete is hereby cancelled, and the belts will continue to be issued for peace requirements.

13853—Rifles, short, magazine, Lee-Enfield   26 Jan 1907
    (Mark III)   C

A pattern of the above-mentioned rifle has been approved to govern future manufacture. It differs from the "Rifles, short, M.L.E. Marks I and I*," described in LoC 11947, 13509, and 13577, in the following particulars—

Blade, foresight— Foresights with a straight edged blade in five heights instead of a barleycorn will be used, according to requirements, to enable variations in shooting to be corrected. They will be marked on the top left side with the figures 0, 015, 03, 045, and 06, respectively, representing 1.0 inch, 1.015 inches, 1.03 inches, 1.045 inches, and 1.06 inches, their heights from the axis of the barrel.

Backsight— The bed is made wider to strengthen the axis joint, and the front part is made tubular so as to encircle the barrel, and is fixed to the barrel by a crosspin and the point of the spring screw to prevent it working loose.

The leaf, which is made to turn over on to the handguard and rebound, is graduated on the top left side by lines representing every 25 yards between 200 and 2,000 yards in addition to the lines on the right for hundreds of yards, as in the Mark I rifle.

The slide may be set at any elevation, above or below the elevation at which the slide was last set, in multiples of 50 yards, by pressing the catch on the left side and releasing a fine-adjustment worm wheel engaged in thread notches on the right side of the leaf, thus enabling the slide to be moved quickly along the leaf. The worm wheel is pivoted in the right side of the catch, and may be rotated in either direction at right angles to the leaf, this movement providing fine-adjustment for the slide. The periphery of the worm wheel is divided by 10 thumb-nail notches, the distance between each notch representing 5 yards in range, i.e., five notches 25 yards, or one division on the left side of the lead, and one complete revolution 50 yards.

The wind gauge is fitted directly on the rear end of the leaf, and is held in position by the wind-gauge screw. A U-notch, instead of a V-notch, is cut in the top edge, and the face is roughened to prevent light being reflected from it. Bright centre lines are marked on the face of the wind gauge for aiming, and on the top as a centre line for use in conjunction with the wind-gauge scale on the leaf.

The wind-gauge scale is marked with divisions representing the same amount of deviation on the target (6 inches per 100 yards) as those on the Mark I rifle. Each quarter turn of the wind-gauge screw represents 1 inch deviation per 100 yards, and at each quarter turn a friction spring engages in a nick inside the head of the screw, thus checking its rotation.

Body— The body is fitted with a bridge charger guide; the slots for the charger stops being sloped in front so that if a charger is left in the charger guide after pressing the cartridges into the magazine, the act of closing the bolt ejects the empty charger.

Bolt head— The charger guide and the slide for the charger guide are omitted.

Butt plate— The trap of the butt-plate is fitted with an axis pin

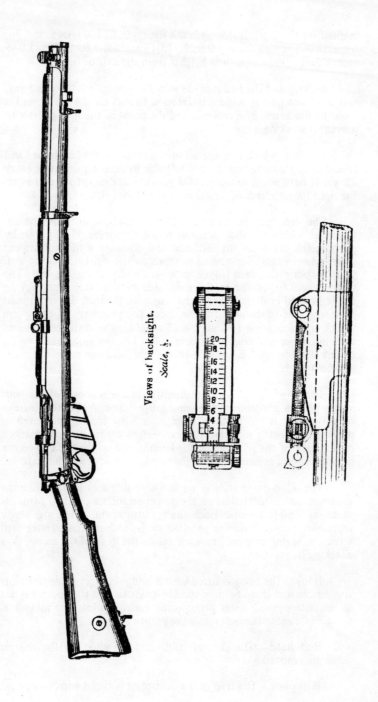

Views of backsight.
Scale, ½.

in a similar manner to that of the "Rifle, M.L.M., Mark II" and the trap spring is tapered in thickness. This butt-plate is similar to the No. 2 butt-plate for Mark I* rifles announced in LoC 13742.

Band, inner— The inner band is 1 inch nearer the breech end of the barrel, and is bored larger to suit the larger diameter of the barrel at that position.

Cut-off— The cut-off is strengthened at the joint end, and is similar to that for Mark I* rifles announced in LoC 13742.

Locking-bolt— The locking bolt is slightly modified to suit the stop pins which have been coned to ensure smoother working of the locking bolt.

Nosecap— The nosecap is lightened and the shape of the fore-sight protecting wings altered to better enable the object to be seen in moving practice, and to admit more light to the foresight.

Handguard, rear— The rear handguard is modified by the removal of the sight protector.

Handguard, front— The front handguard is increased in thickness in front of the outer band to strengthen it.

Stock, fore-end— The position of the inner band seating is moved 1 inch nearer to the breech to strengthen the fore-end; and a new form of backsight protector is fitted to the fore-end and fixed by a screw and nut.

Spring, sight, aperture— The aperture sight spring is recessed for the head of the screw, which is rounded to prevent injury to clothing and accoutrements.

Weight of rifle................. .8 lb. 10½ oz.

The following is a list of the components, showing which are special to this rifle and which are common to other "marks" of short rifles—

Component	Special or Common
Bands—	Common to—
Inner	Marks III and converted IV.
Outer	All Marks.
Blades, foresight	Marks III and converted IV.
Barrel	Marks III and converted IV.
Bead, pointer, dial sight	All Marks.
Bed, backsight	Marks III and converted IV.
Block, band, foresight	All Marks.

Component	Special or Common
Body	Special to Mark III.
Bolts—	Common to—
Breech	All Marks
Locking	Marks III and converted IV.
Stock	Marks I, I*, converted II and III.
Bracket, swivel, butt	Marks I*, converted II*, III and converted IV.
Caps—	
Handguard, front	All Marks.
Nose	Marks III and converted IV, but may be used to replace nosecaps fitted to other Marks.
Case, magazine	All Marks.
Catches—	
Magazine	All Marks.
Safety, bolt, locking	All Marks.
Slide, backsight	Marks III and converted IV.
Clip, stop, magazine	All Marks.
Cocking piece	All Marks.
Collar, screw, front, trigger guard	All Marks.
Cut-off	All Marks.
Disc, marking, butt	Marks I, I*, converted II and III.
Extractor	All Marks.
Guards—	
Hand—	
Front	Marks III and converted IV.
Rear	Marks III and converted IV.
Trigger	All Marks.
Heads—	
Breech bolt	Marks III and converted IV.
Screw, wind gauge	Marks III and converted IV.
Key, block, band, foresight	All Marks.
Leaf, backsight	Marks III and converted IV.
Nut, screw—	
Protector, backsight	Marks III and converted IV.
Back, nosecap	All Marks.
Pins—	
Axis—	
Backsight	Marks III and converted IV.
Trap, butt plate	Marks I, I*, and converted II fitted with No. 2 butt plate, and III.
Worm, fine adjustment	Marks I, I*, and converted II fitted with No. 2 butt plate and III.
Catch, slide, backsight	Marks III and converted IV.
Catch, magazine	All Marks.
Fixing—	
Bed, backsight	Marks III and converted IV.

Component	Special or Common
Pins—	
Fixing—	Common to—
Block, band, foresight	All Marks.
Head, screw, wind gauge	Marks III and converted IV.
Washer, pin, axis, backsight	All Marks.
Joint, band, outer	All Marks.
Screwed, fore-end	Marks I, I* and III.
Stop—	
Band, outer	All Marks. Same as "Pin, stop, band, lower, M.L.M. rifle."
Bolt, locking	Marks III and converted IV.
Trigger	All Marks.
Plates—	
Butt	Mark I* and III, and also to I and converted II when fitted with No. 2 brass buttplates.
Keeper, stockbolt	All Marks.
Sight, dial	All Marks.
Platform, magazine	All Marks.
Pointer, sight, dial	All Marks.
Protector, sight, back	Marks III and converted IV.
Rivets, spring, handguard	All Marks.
Screws—	
Band, inner	All Marks.
Cap, nose—	
Back	All Marks.
Front	All Marks.
Catch, slide, backsight	Marks III and converted IV.
Cut-off	All Marks.
Disc, marking, butt	Marks I, I*, converted II and III.
Ejector	All Marks.
Extractor	All Marks.
Guard, trigger—	
Back	Marks I, I* and III.
Front	All Marks.
Handguard cap	All Marks. Same as "Screw, keeper, fine-adjustment, and handguard cap, &c."
Keeper, striker	All Marks.
Plate, butt	All Marks.
Protector, backsight	Marks III and converted IV.
Sear	All Marks.
Sight, dial—	
Fixing	All Marks.
Pivot	All Marks.
Spring—	
Sight, aperture	Marks I, I* and III.
Trap, butt-plate	Marks I* and III, and I and converted II when fitted with

Component	Special or Common
Screws—	
Spring—	
Sight, back	Common to—
	Marks III and converted IV.
Bracket, swivel, butt	Marks I* and converted II*, III and converted IV. Same as "Screw, strap, buttplate, iron"
Swivel	All Marks.
Wind gauge, backsight	Marks III and converted IV.
Sear	All Marks.
Sight, aperture	All Marks.
Slide, backsight	Marks III and converted IV.
Springs—	
Auxiliary, platform magazine	All Marks fitted with No. 2 magazine.
Catch, slide, sight, back	Marks III and converted IV.
Extractor	All Marks.
Guard, hand, rear, double	All Marks.
Main	All Marks.
Platform, magazine	All Marks.
Retaining breech bolt	All Marks.
Screw—	
Band, inner	All Marks.
Wind gauge, back sight	Marks III and converted IV.
Sear	All Marks.
Sight—	
Aperture	Special to Mark III, but may be used to replace similar spring in Mark I and I*.
Back	Common to—
	Marks III and converted IV, but may be used to replace similar spring in all other Marks.
Dial	All Marks.
Trap, butt-plate	Marks I* and III and I and converted II when fitted with brass buttplates.
Stud, fore-end	All Marks.
Wind gauge, backsight	Marks III and converted IV.
Stocks—	
Butt (long, normal or short)	Marks I* and III.
Fore-end	Special to Mark III, may be used for converted Mark IV if required.
Striker	Common to all Marks.
Studs—	Common to—
Clip, stop, magazine	All Marks.
Fore-end	All Marks fitted with stud and spring.

Component	Special or Common
Swivels—	Common to—
Band, &c.	All Marks.
Piling	All Marks.
Trap, butt-plate	Marks I, I* and converted II fitted with a No. 2 butt-plate, and III.
Trigger	All Marks.
Wad, bolt, stock	All Marks.
Washers—	
Bolt, stock	All Marks.
Guard, hand	All Marks.
Pin, axis, backsight	All Marks.
Screw, fixing, dial sight (and nut, screw, protector, backsight)	All Marks, but in Marks III and converted IV, one additional for "Washer, nut, screw, protector, backsight."
Spring, band, inner	All Marks.
Wind gauge, backsight	Marks III and converted IV.
Worm, fine-adjustment, backsight	Marks III and converted IV.

13854—Rifle, short, magazine, Lee-Enfield, 17 Jun 1907
   converted (Mark IV) C
   Converted from Rifle, M.L.E., Marks I
   and I*; and M.L.M., Marks II and II*.

A pattern of the above-mentioned rifle has been approved to govern future conversion.

This conversion differs from the converted Mark II* described in LoC 13578 in that it embodies the special features of the Mark III described in LoC 13853.

Its weight is 8 lb. 14½ oz., and it is the same as the Mark III arm in all details, the following components excepted—

Component	Remarks
Body	Special to converted Mark IV.
Bolt, stock	Same as M.L.E., No. 2.
Cut-off	Converted as described in LoC 12129 and common to all short rifles; vide LoC 13612.
Disc, marking, butt	Omitted from the converted Mark IV as the butt plate has a strap on which the marking is placed.
Pin, axis, trap, butt plate	Same as M.L.M.
Plate, butt	Same as M.L.M.R., Mark II.
Rivet, fore-end	Same as M.L.M. Not removed from fore-end on conversion.

Component	Remarks
Screw, disc, marking, butt	Omitted from converted Mark IV.
Screw, guard, trigger, back	Same as M.L.M.
Screw, spring, trap, butt-plate	Same as M.L.M.
Screw, spring, sight, aperture	Same as M.L.M.
Screw, strap, butt plate, iron	This screw is also used for "Bracket, swivel, butt," in all short rifles fitted with that component.
Spring, trap, butt plate	Same as M.L.M.
Spring, sight aperture	Same as M.L.M.
Stock, butt	Same as M.L.E.R.S., converted, Mark II*.
Stock, fore-end	Special to converted Mark IV, in that it retains the rivet and two washers on conversion, instead of being fitted with a screwed pin as in the case of the Mark III; but interchangeability of fore-ends is not affected.
Trap, butt plate	Same as M.L.M.R., Mark II.
Washer, rivet, fore-end	Same as M.L.M. Not removed from fore-end on conversion.

13855 – Tube, aiming, .23-inch –   29 May 1907
    Brush, cleaning (Mark II)   C
For all aiming tubes and rifles firing
0.22-inch and 0.23-inch ammunition.

1. New pattern.

    Tube, aiming, .23-inch –
    Rod, cleaning (Mark I)   C
For all aiming tubes and rifles firing
0.22-inch and 0.23-inch ammunition.

2. Change in nomenclature.

    1. A pattern of the above-mentioned brush has been approved to govern future manufacture.

    It differs from that referred to in LoC 7880 as follows –
The shank is made from solid drawn tube, through the top of which the wire is threaded and then twisted to hold the bristles.

    2. Consequent on the introduction of aiming tubes and rifles for firing 0.22-inch ammunition, the detail of the nomenclature of the rod has been altered as shown above.

13856—Tube, aiming, .22-inch, M.L.M. rifle 23 May 1907
   (Mark I) L
   Also Rifles, M.L.E.; for R.F. cartridges, including extractor and screw, set nut and washers.

   Tube, aiming, .22-inch, M.L.M. rifle
   (Mark II) L
   Also Rifles, M.L.E.; for R.F. cartridges; converted from Tube, aiming, M.L.M. rifle, Morris; including extractor and screw, set nut and washer.

Patterns of the above-mentioned tubes have been approved to govern future manufacture and conversion.

None of the Mark I tubes described in LoC 13650 have been made; the Mark I tube described herein has been sealed in substitution, and, consequently, no advance in numeral has been made.

The tubes differ from the converted and Mark I tubes described in LoC 13650 in the following particulars—

   The cartridge head recess of the extractor is continued towards the centre, so that it grips nearly one-half of the circumference of the cartridge head, and thus improves the extraction of the empty cartridge case.

   The rim of the extractor has also been extended to the left, so that when the right side of the rim of the extractor becomes thin by wear, the worn portion of the rim can be filed away by the armourer, and the unworn portion brought into position when the tube is adjusted in the rifle.

Mark II tubes will be marked with the numeral II, on the breechpiece to facilitate identification.

13926—Cartridge, S.A., dummy, drill, .303-inch 31 Jul 1907
   rifles or carbines (Marks I, II and III) C

1. Amplification of instructions for local re-tinning of Marks I and II cartridges.
2. Colouring of wooden bullet of Mark III cartridge.

   1. With reference to the instructions for re-tinning Marks I and II cartridges locally (LoC 12597) it is essential that the tin should be in very small pieces and, accordingly, the words "to be granulated before immersing" should be inserted after "Grain, tin," in list of ingredients. The metal can be thus prepared by pouring the molten tin into a perforated ladle, allowing the small streams of molten metal to drop from a height of 3 or 4 feet into a pail of water. The tin will then be in the form of small, thin, irregular-shaped flakes,

which expose a very large surface to the action of the tinning bath. The cases should be frequently well stirred up with the fragments of tin during the 25 to 30 minutes boiling in process of tinning.

2. In future, the wooden bullet of the Mark III cartridge (LoC 11832, 12597) will be coloured red. Existing stocks will be altered locally by the bullets being immersed in an aniline dye consisting of 6 grains Rose, Bengal, in 3 oz. methylated spirit, care being taken that the wood is well freed from grease before immersion.

13965—Gauge, armourers—   9 Jul 1907
    Distance of bolt from end of chamber,
    .074 inch   **C**
    All magazine rifles and carbines; rejecting gauge.

Approval having been given to increase the rejecting limit of the distance between the face of bolt-head and end of chamber of magazine arms in the hands of troops by .002-inch, a pattern of the above-mentioned gauge has been approved.

Officers commanding units and Ordnance officers in charge of depots, with an armourer attached, and Ordnance officers in possession of the reference gauges referred to in Appendix IV of the Equipment Regulations, Part I, 1906, will demand gauges of the new pattern and return the .072-inch gauges to store.

    For Naval Service the rejecting limit will be as follows—
    For rifles under examination and repair at
        Naval Ordnance depots . . . . .072 inch.
    For rifles on board ships, &c. . . . . .074 inch.

Supply of the gauges to ships, &c., will be arranged by the Admiralty.

13969—Saddlery, universal   1 Nov 1907
    Bucket, rifle, Cavalry, Mark II
    Leather, reversible

A pattern of the above-mentioned bucket has been sealed to govern future manufacture.

It differs from Mark I (LoC 12714) in being made of crop undressed stiff leather, and in being a little larger to take the latest pattern rifle with sling. The sling must be worn attached to the band and butt swivels.

The rifle bucket will be attached to the saddle and surcingle, as shown in drawing in LoC 12714.

    Weight of bucket . . . . .3 lb.

Existing stocks of Mark I buckets will be used up by those units, other than Cavalry, of whose equipment this description of bucket forms a part.

13991—Musket, fencing, short (Mark II)   C   28 Sep 1907
  Converted from Rifles, M.H. With spring bayonet, but without indiarubber pad.

A pattern of the above-mentioned musket has been approved to govern conversion as may be ordered.

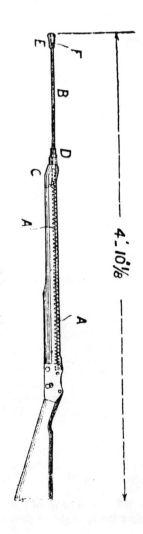

It is similar to the rifle, short, M.L.E., with bayonet, in length, weight and balance. The conversion is made from M.H. rifles, and consists of the following—

The action is removed from the body and the space filled with a wood block.

The barrel is shortened and smoothbored to receive the bayonet spring A and bayonet B, and is fitted with a nosecap C, into which a bayonet stop bush D is screwed. The barrel is hardened and spring tempered.

The bayonet is of steel, hardened and tempered, and is fitted with a steel cap E (instead of an indiarubber pad) which is screwed on the end of the bayonet and fixed by a crosspin F. A balance weight is fitted in the stockbolt hole of the stock butt.

Total length of musket (over all) . . . . .4 ft. 10 1/8-in.
Total weight of musket. . . . . .8 lb. 10 oz.
Balance from end of butt . . . . .24 in.

13992—Rifle, charger-loading, magazine,   1 Jul 1907
    Lee-Metford (Mark II)   L
Converted to charger loading from Rifle, magazine, Lee-Metford, Mark II.

Rifle, charger-loading, magazine,
    Lee-Enfield (Mark I*)   L
Converted to charger loading from Rifles, magazine, Lee-Enfield Marks I and I*, and Lee-Metford, Mark II*.

Patterns of the above-mentioned rifles have been approved to govern conversion.

The conversion consists generally in fitting the rifle with a bridge charger guide and a new magazine for charger loading, an adjustable blade foresight with fixed protector, and a new backsight leaf with a slide having a windgauge, with U notches, adjustable laterally by a screw having its head on the left, and a clamping nut on the right for fixing the slide on the leaf to prevent movement of the slide when firing. The bridge, charger guide and the magazine are similar to those of the "Rifle, short, M.L.E., Mark III" (LoC 13853).

Blade, foresight— Foresights with a straight-edged blade, in five heights, will be used, according to requirements, to enable variations in shooting to be corrected. They will be marked on the left top side with the figures 933, 948, 963, 978, and 993 respectively, representing their heights in decimal parts of an inch, from the axis of the barrel. The blades will be of two patterns— those for the L.M., Mark II (which have the foresight block central) will be marked with the letter M, and those for the L.E. Mark I* (which have the foresight block .027 inch to the left) with the letter E after the figures.

"Rifles, C.L., M.L.E., Mark I*" converted from "Rifles, M.L.M., Mark II*" will be fitted with "blade, foresight" and "protectors, foresight" of the pattern fitted to "Rifles, C.L., M.L.M. Mark II" marked with the letter M, except when fitted with barrels having "Enfield" rifling, which are marked with the letter E on the knoxform. "Rifles, C.L., M.L.M., Mark II" fitted with barrels having "Enfield" rifling, which are marked with the letter E on the knoxform, will be fitted with "blades, foresight" and "protectors, foresight" of the pattern fitted to "Rifles, C.L., M.L.E., Mark I*" marked with the letter E.

Leaf, backsight— The "leaf, backsight" is stronger, and the elevation lines are further from the axis, than in the case of the old leaf removed on conversion, and is without a cap and screw. A stop screw for the slide is fitted at the top corner on the right side.

Slide, backsight— The "slide, backsight" is fitted with a windgauge, adjustable laterally by a screw having a head fixed by a pin on the left side. A clamping nut is fitted to the slide on right side, which, when screwed up, presses a clamping stud on to the edge of the leaf; a friction spring is fitted in the leaf slot on the left side, and presses on the side of the leaf; an extension of this spring acts in the notches cut at right angles on the inside face of the head of the windgauge screw.

The slide is marked with the scale of eight divisions on the front and rear faces, four on the right and four on the left, for the adjustment of the windgauge. Each division represents 6 inches on the target per 100 yards. The slide also has a platinum centre line on the rear face.

The windgauge is fitted into the slide and held and positioned by the windgauge screw, the necessary friction being given by a flat bowed spring fitted in the slide underneath the windgauge. The windgauge has two sighting bars with U notches and platinum centre lines, one for use with leaf down, and one for use when leaf is raised. Lines are marked on the front and rear faces to correspond with the inner line on the slide for centreing the windgauge.

The windgauge screw head is notched at quarter turns to receive the extension to the slide spring. Each quarter-turn represents 1 inch on the target per 100 yards, and six quarter-turns one division of the scale on the slide.

Protectors, foresight— The sight protector is fitted on the barrel and foresight block, and is fixed by a screw. The protectors are of two patterns— one for the "Rifle, C.L., M.L.M., Mark II" marked with the letter M, and one for the "Rifle, C.L., M.L.E., Mark I*" marked with the letter E.

(Note— See "blade, foresight" for remarks as to difference of sight protectors and blades when "Rifles, C.L., M.L.E., Mark I*" are fitted with barrels having "Metford" rifling and "Rifles, C.L., M.L.M., Mark II" are fitted with barrels having "Enfield" rifling.)

Average weight of rifles . . . . .9 lb. 5 oz.

The following is a list of the components, showing which are special to these rifles, and which are common to other arms. Those which are common to other arms are denoted by the word "Common" and their interchangeability is as stated in the "Priced Vocabulary of Stores", &c.—

Component	Special or Common
Bands, lower, M.L.M.R., II	Common
Barrels, M.L.M.C.L.R., II	Special to M.L.M.C.L.R., II, but will be found in M.L.E.C.L.R. I*, that have been converted from M.L.M.R. II*.
Barrels, M.L.E.C.L.R. I* †	Special to M.L.E.C.L.R., I*, but will be found in m.L.M.C.L.R., II that are re-barrelled with barrels having "Enfield" rifling.
Beads, pointer, dial sight, M.L.M.R. II.	Common.
Beds, backsight, M.L.M.C.L.R., II.	Special to M.L.M.C.L.R., II, and M.L.E.C.L.R., I*.
Blades, foresight, M.L.M.C.L.R., II	Special to M.L.M.C.L.R., II, but will be found in M.L.E.C.L.R., I* fitted with barrels having "Metford" rifling.
Blades, foresight, M.L.E.C.L.R., I*	Special to M.L.E.C.L.R., I*, but will be found in M.L.M.C.L.R., II fitted with barrels having "Enfield" rifling.
Bodies, M.L.M.C.L.R. II	Special to M.L.M.C.L.R., II, and M.L.E.C.L.R., I*.
Bolts, breech, M.L.M.C.L.R., II.	Special to M.L.M.C.L.R., II.
Bolts, breech, M.L.E.C.L.R., I*	Special to M.L.E.C.L.R., I*.
Bolts, stock, M.L.E., No. 2	Common.
Caps, nose, M.L.E.R.	Common.
Cases, magazine, M.L.E.R.S., I, No. 2	Common.

† When issued with bodies in exchange for M.L.M.C.L.R., II, the marking must be changed on the lines laid down in LoC 12184.

Catches, magazine, M.L.E.R., II	Common.
Catches, safety, M.L.M.C.	Common (For M.L.E.C.L.R., I* only).
Clips, stop, magazine, M.L.E.R.S., I.	Common.
Cocking pieces, M.L.M.R., II	Common. (For M.L.M.C.L.R., II, only).
Cocking pieces, M.L.M.C.	Common. (For M.L.E.C.L.R., I*, only).
Collars, screw, guard, trigger, front, M.L.M.	Common.
Cut-offs, M.L.E.R.S., I	Common. (Newly manufactured cut-offs, i.e., not converted, are of M.L.E.R.S., I* pattern).
Extractors, M.L.E.R.S., I	Common.
Guards, hand, M.L.M.R., II	Common.
Guards, trigger, M.L.M.C L.R., II	Special to M.L.M.C.L.R., II, and M.L.E.C.L.R., I*.
Heads, breech bolt, M.L.M.C.L.R.,	Special to M.L.M.C.L.R., II, and M.L.E.C.L.R., I*.
Heads, screw, windgauge, backsight, M.L.M.C.L.R., II	Special to M.L.M.C.L.R., II, and M.L.E.C.L.R., I*.
Leaves, backsight, M.L.M.C.L.R. II	Special to M.L.M.C.L.R., II, and M.L.E.C.L.R., I*.
Nut, clamping slide, backsight, M.L.M.C.L.R., II	Special to M.L.M.C.L.R., II, and M.L.E.C.L.R., I*.
Pins, axis, backsight, M.L.M.R.	Common.
Pins, axis, trap, butt plate, M.L.M.	Common.
Pins, catch, magazine, M.L.M.	Common.
Pins, catch, safety, M.L.M.C.	Common. (For M.L.E.C.L.R. I* only).
Pins, fixing, head, screw, windgauge, M.L.M.C.L.R., II	Special to M.L.M.C.L.R., II, and M.L.E.C.L.R., I*.
Pins, stop, band, lower, M.L.M.R.	Common.
Pins, trigger, M.L.M.	Common.
Plates, butt, M.L.M.R., II	Common.
Plates, keeper, stockbolt, M.L.E.R.S., I	Common.
Plates, sight, dial, M.L.M.R., II.	Common.
Platforms, magazine, M.L.E.R.S., I.	Common.
Pointers, sight, dial, M.L.M.R., II	Common.
Protectors, foresight, M.L.M. C.L.R., II	Special to M.L.C.L.R. II, will be found on M.L.E.C.L.R., I*, fitted with barrels having "Metford" rifling.
Protectors, foresight, M.L.E. C.L.R., I*	Special to M.L.E.C.L.R., I*, will be found on M.L.M.C.L.R., II, fitted with barrels having "Enfield" rifling.

Rivets, fore-end, M.L.M.	Common.
Rivets, spring, handguard, M.L.M.	Common.
Screws, band, lower, M.L.M.R., II	Common.
Screws, bed, backsight, M.L.M.R.	Common.
Screws, cap, nose, M.L.M.R.	Common.
Screws, cut-off, M.L.M.	Common.
Screws, ejector, M.L.M.	Common.
Screws, extractor, M.L.M.	Common.
Screws, guard, trigger, M.L.M., back.	Common.
Screws, guard, trigger, M.L.M., front.	Common.
Screws, keeper, striker, M.L.M.R.	Common. (For M.L.M.C.L.R., II, only).
Screws, keeper, striker, M.L.M.C.	Common. (For M.L.E.C.L.R., I*, only).
Screws, plate, butt, iron	Common.
Screws, protectors, foresight, M.L.M.C.L.R., II	Special to M.L.M.C.L.R., II, and M.L.E.C.L.R., I*.
Screws, sear, M.L.M.	Common.
Screws, sight, aperture, M.L.M.R.	Common.
Screws, sight, dial, M.L.M., fixing	Common.
Screws, sight, dial, M.L.M., pivot	Common.
Screws, spring, sight, aperture, M.L.M.	Common.
Screws, spring, sight, back, M.L.M.	Common.
Screws, spring, trap, butt plate, M.L.M.	Common.
Screws, stop, slide, backsight, M.L.M.C.L.R., II	Special to M.L.M.C.L.R., II, and M.L.E.C.L.R., I*.
Screws, strap, butt plate, iron	Common.
Screws, swivel, piling, M.L.M.R., II.	Common.
Screws, windgauge, M.L.M.C.L.R. II	Special to M.L.M.C.L.R., II, and M.L.E.C.L.R., I*.
Sears, M.L.M.R., II	Common.
Sights, aperture, M.L.M.R.	Common.
Slides, backsight, M.L.M.C.L.R. II	Special to M.L.M.C.L.R., II, and M.L.E.C.L.R., I*.
Springs, auxiliary, platform, magazine, M.L.E.R.S., I, No. 2	Common.
Springs, extractor, M.L.M.	Common.
Springs, guard, hand, M.L.M.R., II, lower	Common.
Springs, guard, hand, M.L.M.R., II, upper	Common.
Springs, main, M.L.M.	Common.
Springs, pin, catch, safety, M.L.M.C.	Common. (For M.L.E.C.L.R., I*, only).

Springs, platform, magazine, M.L.E.R.S., I	Common.
Springs, retaining, breech bolt, M.L.M.	Common.
Springs, sear, M.L.M.	Common.
Springs, sight, aperture, M.L.M.	Common.
Springs, sight, back, M.L.M.R.	Common.
Springs, sight, dial, M.L.M.	Common.
Springs, slide, backsight, M.L.M. C.L.R., II	Special to M.L.M.C.L.R., II, and and M.L.E.C.L.R., I*.
Springs, trap, butt plate, M.L.M.	Common.
Springs, windgauge, backsight, M.L.M.C.L.R., II	Special to M.L.M.C.L.R., II, and M.L.E.C.L.R. I*.
Stocks, butt, M.L.M.R., II, No. 2 (long, normal or short)	Common.
Stocks, fore-end, M.L.M.C.L.R., II	Special to M.L.M.C.L.R., II, and M.L.E.C.L.R., I*.
Strikers, M.L.M.R.	Common. (For M.L.M.C.L.R., II, only).
Strikers, M.L.M.C.	Common. (For M.L.E.C.L.R., I*, only).
Studs, clamping, slide, backsight, M.L.M.C.L.R., II	Special to M.L.M.C.L.R., II, and M.L.E.C.L.R., I*.
Studs, clip, stop, magazine, M.L.E.R.S., I	Common.
Swivels, band, M.L.M.R., II	Common.
Swivels, butt, M.L.M.R.	Common.
Swivels, piling, M.L.M.R., II	Common.
Traps, butt plate, M.L.M.R. II	Common.
Triggers, M.L.M.R., II	Common.
Wads, bolt, stock, M.L.M.	Common.
Washers, bolt, stock	Common.
Washers, guard, hand, or rivet, fore-end, M.L.M.	Common.
Washers, screw, fixing, dial sight, M.L.M.R.	Common.
Windgauges, backsight, M.L.M. C.L.R., II	Special to M.L.M.C.L.R., II, and M.L.E.C.L.R., I*.

13993—Sword, practice, gymnasia, pattern 1907 (Mark I)   L   6 Nov 1907
Without scabbard.

A pattern of the above-mentioned sword has been approved to govern manufacture.

It differs from the "Sword, practice, gymnasia, pattern 1904, Mark I" (LoC 12328), in the following particulars—

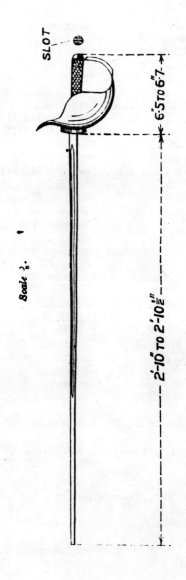

The blade is 1.2 inches longer, and the grip is made of aluminium or aluminium alloy. The blade is screwed into the pommel, on the end of which a slot .07 inch wide and .125 inch deep is cut to enable it to be turned with a screw-driver or coin when fitting a new blade.

The first supplies of these swords (described as "Mark IA", but which should now be described as above) were not provided with this slot, which should be cut locally by armourers.

13994—Tools, sight line—   12 Oct 1907
    Rifle, short, M.L.E., No. 2   C
    Steel; reversible. For Rifle, short, M.L.E.,
    Mark III, and converted, Mark IV.

    Plates, screw—
      6 taps, 6 holes—   C
    With handles. For special threads of
      screws, Rifles, short, M.L.E.
      Taps—
        No. 11 plug   C
        No. 11 taper   C
        No. 12 plug   C
        No. 12 taper   C
        No. 13 plug   C
        No. 13 taper   C

1. New patterns.

    Tools, sight line—
    Rifle, short, M.L.E., No. 1   C
    Steel; reversible. For Rifles, short, M.L.E.,
    Marks I and I*, and converted, Marks II and II*.

2. Change in nomenclature.

1. Patterns of the above-mentioned tools have been approved to govern manufacture.

The sight line is for re-lining the centre lines on the windgauge of the backsight of "Rifles, short, M.L.E., Mark III, and converted, Mark IV". It is reversible, so that the lining may be done from both sides of the sight leaf to ensure central alignment, as in the case of the tool described in LoC 12422.

When re-lining the windgauge, care should be taken to adjust it until the original centre line, cut into the chequered face, is beside the front lining edge of the tool, and the centre line on top of the sight leaf beside the top lining edge of the tool. The scriber described in LoC 12422 will be used with this tool.

The screw plate and taps are required for the special threads of screws of "Rifles, short, M.L.E." the "Plate, screw, 20 taps, 20 holes" described in LoC 7323, being suitable for the remaining threads.

The diameter and pitch of threads of the holes in plate and of the taps, and the screws for which they are suitable, is as follows—

	Holes in plate and taps.			Number of screws per rifle.					
				Rifles, short, M.L.E.					
			Screws.				Converted.		
Number.	Diameter of thread.	Number of threads per inch.		Mark I.	Mark I.*	Mark III.	Mark II.	Mark II.*	Mark IV.
	in.								
11	·096	56	Keeper, line adjustment, sight, back; handguard cap; and sight protector top.	4	4	2	4	4	2
12	·125	20†	Windgauge	1	1	..	1	1	..
13	·17	46	Windgauge	..	..	1	..	..	1

† Double Thread.

2. In consequence of the adoption of the No. 2 sight line tool for "Rifles, short, M.L.E., Mark III, and converted, Mark IV" the nomenclature of the tool described in LoC 12422 has been amended to read as shown above.

14042—Rifles, short, M.L.E., A.T. (Mark I)   L   11 Mar 1907
   For .22-inch rimfire cartridge aiming tube.   30 Dec 1907

Rifles, short, M.L.E., $\frac{A.T.}{N.I.}$ (Mark I)   L

   For .22-inch rimfire cartridge aiming tube; non-interchangeable; part of 1,000 "Rifles, short, M.L.E.," that were issued for trial.

1. Not to be used for drill purposes.
2. Nomenclature altered.

With reference to LoC 13649—
1. The above-mentioned rifles will be used for aiming-tube practice only, and "D.P." rifles will be used in future for drill purpose.

2. The detail of the nomenclature of the above-mentioned rifles has been amended to read as above.

14043—Tool, removing hand-guard, No. 1 C   16 Jan 1908
   For Rifles, M.L.M., M.L.E. and M.E., and
   Rifles, short, M.L.E., Mark III, and converted,
   Mark IV.

1. Use extended to Rifles, short, M.L.E., Mark III, and converted, Mark IV.
2. Change in nomenclature.

The use of the above-mentioned tool (LoC 6379), described in "Priced Vocabulary of Stores, 1906" as pattern A, having been extended as shown above, the nomenclature has been amended accordingly.

The "Tool, removing handguard, No. 2" (LoC 11079) is not suitable for short rifles of any pattern, but will still supersede the No. 1 pattern for use with M.L.M., M.L.E. and M.E. rifles.

14069—Guard, hand, front, M.L.M. rifle   21 Oct 1907
   Asbestos; Patent Nos. 13754 of 1899, and
   6934 of 1907; also for M.L.E. and M.L.M.C.L.
   and M.L.E.C.L. rifles.

A pattern of the above-mentioned handguard has been sealed to govern supplies for the use of officers, non-commissioned officers and men in voluntary practices.

The handguard is made of unglazed black cotton twill, with a wire frame sewn in, and lined with asbestos; five bands of black webbing are sewn on to fix the handguard to the rifle, the bands being held by spring buttons. The handguard is fitted on the rifle between the front and back sights, and completely covers the exposed portion of the barrel.

The handguard will not be supplied from store, and when required is to be obtained direct from the trade by the purchaser.

Illustration of LoC 14069, overleaf.

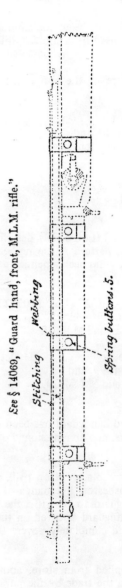

See § 14069, "Guard hand, front, M.L.M. rifle."

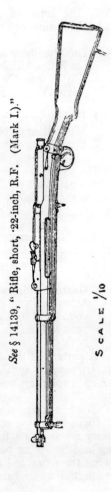

See § 14139, "Rifle, short, ·22-inch, R.F. (Mark I.)."

SCALE 1/10

14070—Rifles, short, M.L.E.   C   21 Oct 1907
All marks   23 Nov 1907

Fore-end and nose-cap to be numbered to correspond with barrel number.

As it is essential to good shooting that each rifle should be used with the fore-end and nose-cap which were assembled to it when sighting was adjusted, the fore-ends and nose-caps of the above-mentioned rifles in store and in the hands of troops, either in peace or in mobilization equipment, will be numbered by armourer serjeants to correspond with the barrel number.

In the case of units not provided with an armourer-serjeant, the General Officer Commanding will arrange for the service to be performed either by regimental or A.O.D. armourers, whichever is the more economical course. Reservists arms will be numbered at the same time as the arms of the regimental depot.

The fore-end will be marked underneath, around the radius, about 1 inch in rear of the nose-cap, with "stamps, steel, 1/8-inch for metal" (or "for wood" is available)— figures and letters; and the nose-cap, on the end of the sword-bayonet boss, with "stamps, steel, 5/64-inch for metal"— figures and letters.

Should a spare fore-end or spare nose-cap subsequently be fitted to a rifle, it will be similarly marked after the rifle has passed the firing test prescribed in the Instructions for Armourers, 1904. The fore-ends and nose-caps of existing rifles in Naval Service will be numbered locally at depots.

14131—Cartridge, S.A. ball, .303-inch, cordite   C   11 Apr 1907
                                                          1 Oct 1907
Alteration of marking on base.   16 Oct 1907
                                                         25 Oct 1907

The above-mentioned cartridge will in future have the last two figures of year of manufacture stamped on the base, the year, for convenience of manufacture, dating from 1st April to 31st March.

The letter "C" denoting cordite, will be omitted.

Certain issues of cartridges embodying the above alterations have already been made.

14139—Rifle, short, .22-inch, R.F. (Mark I)   L   13 Dec 1907
Converted from "Rifle, M.L.M., Mark I*"

A pattern of the above-mentioned rifle has been approved to guide conversion as may be ordered.

The conversion is made from "Rifle, M.L.M., Mark I*", and consists generally in altering the rifles to approximately as nearly as possible in weight, length, &c., to the short rifle, and in fitting a new barrel with back and front sights similar to those of the "Rifle, short, M.L.E., Mark III" described in LoC 13853, but bored and chambered to suit the "Cartridge, aiming tube, R.F., Mark I" (LoC 13642). The bolt action is converted, a suitable bolt-head and striker being fitted.

Body— The gas vents left and right of body are enlarged.

Bolt, breech— The bolt has the cover stops and a portion of the top of the rib removed, and the cover screw holes filled in. The front end is bored to receive a threaded bush, which is fixed by a portion of the original bolt-head screw, the bush and screw being brazed in. The bush is tapped to receive the screwed tenon of the bolt-head.

Bolt-head— The striker hole in the face of the bolt-head is bored .0885-inch eccentric to the axis of the bolt-head, and the hole for the extractor axis screw is bored nearer the axis of the bolt-head than in bolt-heads of "Rifles, short, M.L.E., A.T., Mark I" described in LoC 13649. For the purpose of identification, bolt-heads of this pattern are marked ".22" on the top of the wing.

Striker— The striker is arranged suitably for the bolt-head, having a flat on the front portion to clear the rear end of the extractor.

Stock, fore-end— The stock, fore-end is cut short, and a liner is fitted, glued in, and grooved suitably for the smaller barrel. The dial sight seating and the clearing rod groove are filled up.

Extractor— An extractor of suitable form, having a long and narrow hook, is fitted. It is so arranged that the outward movement is limited by the body to prevent injury to the extractor spring in the event of a burst cartridge case.

Bands (upper and lower)— The upper and lower bands have liners, suitable for the smaller barrel, fitted and brazed in.

Nosecap— The bar for the sword-bayonet is removed, and the barrel groove has a liner, suitable for a smaller barrel, fitted and brazed in.

Protector, foresight— A foresight protector is fitted on the barrel, over the foresight block, and is fixed by a screw.

Particulars relating to rifling, sighting, weight, &c.

Length of barrel . . . . .25 3/16 in.
Calibre . . . .214 inch.
Rifling—
  Grooves, number. 8
  Grooves, depth (mean) . . . .0052 inch.
  Grooves, width . . . .05 inch.
  Width of lands. . . . .03 inch.
  Twist of rifling . . . . .Right hand,
                                      1 turn in 16 ins.
Sighting system—   Adjustable blade foresight, radial backsight, with fine adjustment and windgauge, as on "Rifle, short, M.L.E., Mark III."
Distance between blade, foresight, and backsight, U . . . . .19 15/32-inches.
Length of rifle .3 feet 8½ inches.
Weight of rifle. .8 lb. 4 oz.
Ammunition: "Cartridge, aiming tube, R.F. (Mark I) (LoC 13642.

The following are the approximate elevations required at different ranges—

Range	*Elevation of backsight*
yards.	yards.
25	300
50	350
100	450
150	550
200	650

The rod and brush, described in LoC 7880 and 13855, are suitable for this rifle, and should be inserted from the breech end. Under no circumstances should the rod and brush be inserted from the muzzle for cleaning the barrel, as the friction of the rod is liable to enlarge the bore and make the barrel bell-mouthed.

The following is a list of components showing which are special to this rifle, and which are common to other arms. Those which are common to other arms are denoted by the word "Common" and their interchangeability is as stated in the "Priced Vocabulary of Stores" &c.

Component	*Special or common*
Bands—	
Lower, R.F.R.S., I	Special. Converted from M.L.M.R., I*.
Upper, R.F.R.S., I	Special. Converted from M.L.M.R., I*.
Barrels, R.F.R.S., I	Special.

Component	Special or common
Beds, backsight, M.L.E.R.S., III	Common.
Blades, foresight, M.L.E.R.S., III	Common.
Blocks, band, foresight, M.L.E.R.S., I	Common.
Bodies, R.F.R.S., I	Special. Converted from M.L.M.R., I*.
Bolts—	
Breech, M.L.M.R., I*, R.F.	Common. Converted from M.L.M.R., I*.
Stock, M.L.E., No. 1	Common.
Caps, nose, R.F.R.S., I	Special. Converted from M.L.M.R., I*.
Catches—	
Magazine, M.L.M.R., I*	Common.
Slide, backsight, M.L.E.R.S., III	Common.
Cocking pieces, M.L.M.R., I*	Common.
Collars, screw, front, trigger guard, M.L.M.	Common.
Discs, marking, butt	Common.
Extractors, R.F.R.S., I	Special.
Guards—	
Hand, R.F.R.S., I	Special. Converted from "Guard, hand, rear, M.L.E.R.S., I."
Trigger, M.L.M.R., I*	Common.
Heads—	
Breech bolt, R.F.R.S., I	Special.
Screw, wind-gauge, M.L.E.R.S., III	Common.
Keys, block, band, foresight, M.L.E.R.S., I	Common.
Leaves, backsight, M.L.E.R.S., III	Common.
Nuts, screw, band, upper, M.L.M.R.	Common.
Pins—	
Axis—	
Backsight, M.L.E.R.S., III	Common.
Trap, butt-plate, M.L.M.	Common.
Worm, fine-adjustment, catch, slide, backsight, M.L.E.R.S., III	Common.
Catch, magazine, M.L.M.	Common.
Fixing—	
Bed, backsight, M.L.E.R.S., III	Common.
Block; band, foresight, M.L.E.R.S., III	Common.
Head, screw, wind-gauge, M.L.E.R.S., III	Common.
Washer, pin, axis, backsight, M.L.E.R.S., I	Common.
Stop, band—	
Lower, M.L.M.R.	Common.
Upper, M.L.M.R.	Common.

Component	Special or common
Pins—	
Trigger, M.L.M.	Common.
Plates, butt, M.L.M.R., I*	Common.
Protectors—	
Foresight, R.F.R.S., I	Special.
Sight, handguard, rear, M.L.E.R.S., I	Common.
Rivets—	
Fore-end, M.L.M.	Common.
Spring, handguard, rear, M.L.E.R.S., I (2)	Common.
Screws—	
Band—	
Lower, M.L.M.R., I*	Common.
Upper, M.L.M.R.	Common.
Cap, nose, M.L.M.R.	Common.
Catch, slide, backsight, M.L.E.R.S., III	Common.
Disc, marking, butt	Common.
Ejector, M.L.M.	Common.
Extractor, M.L.M.	Common.
Guard, trigger—	
M.L.M., Back	Common.
M.L.M., Front	Common.
Keeper—	
Fine adjustment; handguard cap; and sight-protector, top; M.L.E.R.S., I	Common. (For sight protector top only).
Striker, M.L.M.R.	Common.
Plate, butt, iron (2)	Common.
Protector—	
Foresight, M.L.M.C.L.R., II	Common.
Sight, side, handguard, rear, M.L.E.R.S., I	Common.
Sear, M.L.M.	Common.
Spring—	
Sight, back, M.L.E.R.S., III	Common.
Trap, butt plate, M.L.M.	Common.
Swivel, trigger guard, M.L.M.R.	Common.
Wind-gauge, backsight, M.L.E.R.S., III	Common.
Sears, M.L.M.R., I*	Common.
Slides, backsight, M.L.E.R.S., III	Common.
Springs—	
Catch, slide, backsight, M.L.E.R.S., III	Common.
Extractor, M.L.M.	Common.
Guard, hand, rear, double, M.L.E.R.S., I	Common.

Component	Special or common
Springs—	
Main, M.L.M.	Common.
Retaining breech-bolt, M.L.M.	Common.
Screw, wind-gauge, backsight, M.L.E.R.S., III	Common.
Sear, M.L.M.	Common.
Sight, back, M.L.E.R.S., III	Common.
Trap, butt-plate, M.L.M.	Common.
Wind-gauge, backsight, M.L.E.R.S., III	Common.
Stocks—	
Butt, M.L.M.R., I	Common.
Fore-end, R.F.R.S., I	Special. Converted from M.L.M.R., I*.
Strikers, R.F.R.S., I	Special. Suitable to replace "Strikers, M.L.M.R., I*, R.F."
Swivels—	
Band, M.L.M.R., I*	Common.
Butt, M.L.M.R.	Common.
Guard, trigger, M.L.M.R.	Common.
Piling, M.L.M.R., I*	Common.
Traps, butt-plate, M.L.M.R., I*	Common.
Triggers, M.L.M.R., I*	Common.
Washers—	
Bolt, stock	Common.
Guard, hand, or rivet, fore-end— M.L.M. (2)	Common. For "Rivet, fore-end."
M.L.E.R.S., I (2)	Common. For "Rivet, fore-end."
Pin, axis, backsight, M.L.E.R.S., I	Common.
Wind-gauges, backsight, M.L.E.R.S., III	Common.
Worms, fine-adjustment, backsight, M.L.E.R.S., III	Common.

14168—Rifles—   19 Jul 1907
    M.L.M., A.T.   L
    M.L.E., A.T.   L
    For .22-inch rimfire cartridge, aiming tube practice.

    Rifles described as above have been approved for use in the Territorial Force (Infantry and Engineers) for aiming tube practice.

    The rifles, according to pattern, will be fitted with one of the .22-inch R.F. breech bolts described in LoC 13650, and the chamber of the barrels bushed so that Service ball or blank ammunition cannot be fired from them.

These rifles will be marked A.T. (aiming tube) on the body, barrel, bolt, and stocks, butt and fore-end, and classified as such.

14169—Sword-bayonet, pattern 1907 (Mark I) C 30 Jan 1908
For rifles, short, M.L.E. 15 Apr 1908

Scabbard, sword-bayonet, pattern 1907 (Mark I)
Brown leather.

Patterns of the above-mentioned sword-bayonet and scabbard have been approved for future manufacture.

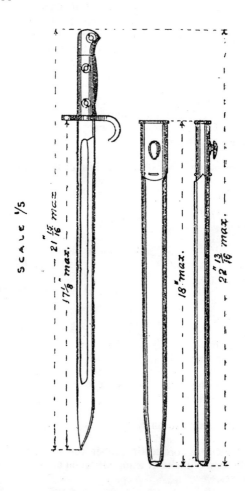

Sword-bayonet.

The sword bayonet differs from the pattern 1903, described in LoC 11716 in the following particulars—

The blade is 5 inches longer and single edged, and has a fuller (or lightening groove) on both sides. One end of the crosspiece is formed as a hook.

Weight of sword-bayonet . . . . .16 1/2 oz.
Length of sword-bayonet blade, maximum. .17 1/8 inches.
Length of sword-bayonet overall . . . . . . .21 15/16 inches.

Scabbard.

The scabbard differs from the pattern 1903, Land, Mark I described in LoC 11811 in the following particulars—

It is made of a shape and length suitable for the longer blade of the sword-bayonet. The leather tag is omitted and a stud is fitted and brazed on to the locket for fastening the scabbard in the new web frog (not yet published in List of Changes) which is hung on the waistbelt.

Weight of scabbard . . . . .7 oz.
Length of scabbard, maximum. . . . . .18 inches.
Length of sword-bayonet and scabbard,
  overall, maximum . . . . .22 13/16 inches.

The following table shows the interchangeability of the components—

Component	Remarks
Sword-bayonet—	
Bolt	Same as pattern 1903.
Grip (pair)	Special.
Nut, bolt	Same as pattern 1887, Mark II.
Nut, screw, grip (2)	Same as pattern 1887, Mark III.
Screw, grip (2)	Same as pattern 1887, Mark III.
Spring, bolt, spiral	Same as pattern 1887, Mark III.
Scabbard—	
Chape	Same as pattern 1903, Land, Mark I.
Lace, iron, short (2)	Common.
Leather, finished	Special.
Locket	Special.
Rivet, spring, locket (6)	Same as pattern 1888.
Spring, locket (2)	Special.

14204—Boxes, ammunition, S.A.   C   1 Jan 1908
                                                                  6 Feb 1908
  1. Omission of cordite lot number from markings.   7 Apr 1908
  2. Marking of make of ammunition on ends of
     boxes.

1. With reference to LoC 10098, the lot number of cordite will, in future, be omitted from the label on lining, and from the stencilled markings on the above-mentioned boxes.

2. In order to facilitate identification of the make of ammunition when the boxes are stacked, the initial, or initials, of the manufacturer of the cartridges will be stencilled on the ends of the boxes, in addition to the date.

14207—Action, skeleton, M.L.E., short (Mark III)   C   21 May 1908
   Also Rifle, short, M.L.E., converted, Mark IV.

A pattern of the above-mentioned skeleton action has been sealed to govern future manufacture.

This action differs from the "Action, skeleton, rifle, short, M.L.E., Mark I" (LoC 12220) in being fitted with a bridge charger guide similarly to the "Rifle, short, M.L.E., Mark III" (LoC 13853).

14209—Oil, petroleum, Russian, lubricating   17 Jul 1908

1. Extended use.

Mineral jelly, yellow

2. Use discontinued.

1. Oil, petroleum, Russian, lubricating, will in future be the only preparation to be used for cleaning and oiling the actions and bores of .303-inch rifles, carbines and machine guns in the hands of troops. It will also constitute the sole lubricant and preservative for bicycles. The use of Rangoon oil for these purposes will therefore be discontinued.

2. Mineral jelly, yellow, being no longer required for the above purposes, and its use as footgrease having been discontinued, will be considered obsolete as soon as existing stocks have been used up.

14232—Box, ammunition, small-arm, 1,000 rounds,   19 Apr 1907
   .303-inch, in chargers, No. 2 (Mark I)   L

Obsolete.

   Box, ammunition, small-arm, 1,000 rounds,
   .303-inch, in chargers (Mark I)   L
   Wood, with tin lining.

Nomenclature.

With reference to LoC 11980, 12157— Consequent upon the

No. 1 small-arm ammunition box, which is made of soft wood, being made suitable for use in foreign climates by being treated with oil, mineral, preserving wood (LoC 14207), no more of the No. 2 boxes will be made, and so soon as the existing stock is used up they will become obsolete.

The nomenclature and details appended thereto of the No. 1 box have therefore been amended to read as shown above.

14234—Carbines—   25 Jun 1908
    M.L.M. and M.L.E.
    M.M. and M.E.

    Rifles—
    M.L.M. and M.L.E.
    M.M. and M.E.

Shortening of lead from chamber to bore.

In order to better centre the bullet in the bore, and to increase the life of the barrel, it has been decided for future manufacture of barrels of the above arms to reduce the length of the lead from chamber to bore from .8-inch to .6-inch, similar to that of Rifles, short, M.L.E.

Barrels with the .6-inch lead will not be fitted to carbines or rifles on payment until the stock of new barrels fitted with .8-inch lead is used up.

14236—Rifles, short, M.L.E.—   9 Apr 1908
    Mark III   C
    Converted, Mark IV   C

    Rifles, charger-loading—
    M.L.M., Mark II   L
    M.L.E., Mark I*   L

Heavier extractor spring.

It having been found necessary to increase the weight of the extractor spring in the above-mentioned rifles from, "from 4½ to 5½ lb." to "from 7 to 9 lb.", LoC 13853 & 13992 will be amended as follows—

    LoC 13853—
        For "Springs, extractor   All Marks"
        Read "Springs, extractor   Common to Marks III
                                          and converted IV, but
                                          may be used to replace
                                          similar spring in all other
                                          marks."

LoC 13922–
For "Springs, extractor, M.L.M."
Read "Springs, extractor, M.L.E.R.S. III."

For purpose of identification this spring will be marked with a figure 3 inside the long end.

For Naval Service– Arrangements will be made by Admiralty for the light extractor springs in naval rifles to be exchanged before the rifles are issued to ships.

14288–Web equipment, pattern 1908–   31 Jan 1908
   Belt, waist–
     Large
     Small
   Brace with buckle
   Carriers, cartridge, 75 rounds–
     Left
     Right
   Carrier, water-bottle
   Frog
   Haversack
   Pack
   Strap, supporting

New patterns.

Patterns of the above-mentioned articles have been sealed to govern manufacture.

The equipment is made throughout of woven web, all the buckles used being of the self-locking or tongueless pattern.

The waistbelt is in two sizes, the overall length being 48 and 40 inches respectively, the width in each case being 3 inches. It is fitted with a large buckle in front, two smaller buckles and two end pieces in the centre of the back. The length of the belt is adjustable at the buckle end.

The braces are issued two to a set, and are interchangeable; each consists of a strip of webbing 50 inches long and 2 inches wide. The width is increased to 3½ inches for a short distance about the middle of the brace. Each brace is provided with a sliding buckle, for the attachment of the pack.

Cartridge carriers– These consist of an assemblage of five 15-round pockets in two tiers, arranged for left or right side, and marked "L" and "R" respectively, each pocket being provided with a separate cover and secured by means of snap fasteners.

The carrier is fitted with a double hook at each end for attaching it to the waistbelt, to which it is further secured by two narrow bands passing round the belt and snapped on to studs on the lower front edge of the carrier.

Water-bottle carrier— This is a skeleton framework of webbing in which the water-bottle is placed and secured by a snapped-on retaining strap. The carrier is fitted with two buckles for attachment to the end pieces of the equipment, and has a short extension piece and snap fastener for use when being carried on the front of the haversack.

The frog consists of a loop to slip on the waistbelt and a body fitted with two loops.

The haversack is a rectangular bag, approximately 11 inches by 9 inches by 2 inches and has a flap cover secured by two small straps and buckles. It is also fitted on the back with two suspending tabs, at the ends with large buckles and on the front with smaller buckles at the bottom corners.

Another small buckle is fitted to the top of the cover, and a stud is placed lower down on the haversack to allow of the water-bottle being carried thereon, if necessary.

The pack is a rectangular-shaped sack, approximately 15 inches by 13 inches by 4 inches and closed at the top by a folding cover secured by two narrow straps.

A short suspension tab is fixed to each of the upper corners at the back, also two small buckles to which are attached the upper ends of the supporting straps, and there are two web loops at the bottom through which the supporting straps are passed.

The supporting straps are issued two in a set; they are interchangeable, and each consists of a strip of 1-inch webbing 31½ inches long, fitted with a buckle at one end.

14289—Web equipment, pattern 1908—   27 Apr 1908
Haversack, converted.

A sample of the above-mentioned article has been sealed to guide conversion of "Haversacks, G.S." as ordered.

The conversion consists in the removal of the slings, and fitting the haversack with two side chapes and buckles, two suspending tabs, also three small chapes and buckles and bottom part of snap fastener.

All the buckles are of the tongueless pattern, the suspending tabs and small chapes of woven web, and the side chapes are made from the old slings.

14319—Cartridge, S.A., ball, .303-inch cordite   C   10 Apr 1907

Stamping of broad arrow on the base.

Contract supplies of the above-mentioned cartridges will, in future, have two broad arrows stamped on the base. All cartridges issued bearing "08" on the base, have this additional stamping.

Cartridges of R.L. manufacture will, as hitherto, be stamped R ↑ L.

14325—Sword, cavalry, pattern 1908 (Mark I)   L   6 Jul 1908
With buff piece; steel hilt.

Scabbard, sword, cavalry, pattern 1908 (Mark I)   L
Steel; with wood lining.

Patterns of the above-mentioned sword and scabbard have been approved for future manufacture.

Sword

The sword differs from cavalry swords of previous pattern mainly in being fitted with a straight tapering blade to facilitate thrusting. The guard is shaped to afford more protection on the left side; the outer edge is beaded. The grip is in one piece and is formed to fit the hand; a recess for the thumb is cut in the back at the front end. The tang is tapered and passes through the guard, ferrule, and grip, which are held together by a nut screwed on the tang under the pommel. The pommel, also, is held by a nut screwed on the tang and the end of the tang is lightly riveted over the nut.

A buff leather washer is cut to the size of the mouth of the scabbard, and waterproofed by soaking in paraffin wax, is placed on the blade next the guard.

Weight of sword complete— 2 lb. 13½ oz. to 2 lb. 15½ oz.
Balance—   2½ to 3 inches from hilt.

Scabbard.

The scabbard of steel is a straight taper from end to end, having a tip, instead of a shoe, brazed in at the bottom end, and a mouthpiece fitted and secured to the top by two screws. Two loops are fitted opposite each other at the top end, and a solid wood lining is fitted inside the scabbard.

Weight of scabbard, complete— 1 lb. 5 oz. to 1 lb. 7 oz.

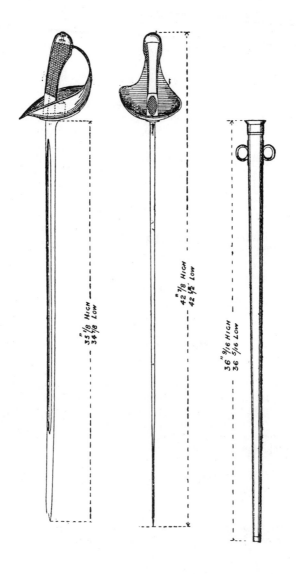

The screws, mouthpiece, of the above-mentioned scabbard (which are common to all cavalry sword scabbards) are the only components of the above-mentioned sword and scabbard common to other patterns.

14326—Tools, adjusting foresight, rifle,   6 Aug 1908
    charger-loading, M.L.M.—
    And M.L.E.
      Cramp (Mark I)   L
      Steel; with two adjusting screws.

1. New Pattern.

    Tools, adjusting foresight, rifle,
    short, M.L.E.—
      Punch, centre (Mark I)   C
      Steel; and for rifles, charger-
      loading, M.L.M. and M.L.E.

2. Change in nomenclature.

    1. A pattern of the above-mentioned cramp has been sealed to govern manufacture.

    It differs from the cramp for Rifles, short, M.L.E. (LoC 12993) principally in being made suitable for the lower position of the foresight blade of charger-loading rifles. It will be used similarly to that for Rifles, short, M.L.E., after having first removed the foresight protector and screw.

    2. As the centre punch for Rifles, short, M.L.E., (LoC 12993) is also suitable for use with charger-loading rifles, the details of its nomenclature has been amended to read as shown above.

14362—Rifles, short, magazine, Lee-Enfield Mark I.   21 Jan 1908
    Rifles, short, magazine, Lee-Enfield
    converted, Mark II.

Fitting of butt plate with trap.

    The steel butt plates of the above-mentioned rifles in the Land Service will be gradually exchanged for the Mark I* rifle, short, M.L.E., brass butt-plate, with trap. Instructions as to the order in which the work is to proceed will be issued in due course. The butt plates will be fitted locally by Regimental and Army Ordnance Department Armourers.

    The following special tools will be supplied for this purpose—
Blade, saw, keyhole, 9-inch— For fitting into "handle, file, large"
Tool, recessing butt plate clearance.

Instructions for fitting Plates, butt, rifle, short, M.L.E., Mark I,
to rifles, short, M.L.E., Marks I and converted Mark II.*

Remove the butt plate from the butt of the rifle, short, M.L.E.

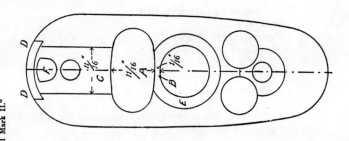

See § 14362, "Rifles, short, magazine, Lee-Enfield, Mark I., and converted Mark II."

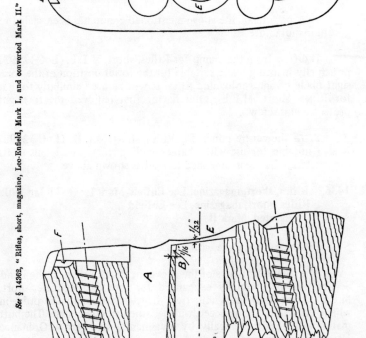

Mark I, or converted Mark II, and alter the butt plate seating of the stock butt, as shown in the accompanying drawing.

With the recessing tool cut a recess around the stockbolt hole for butt plate clearance at E, to a depth of 1/32-inch, full at the shallowest part.

With the keyhole saw cut away the wood between the two upper lightening holes at A for the pull-through weight, at the same time increasing the width of the opening in the narrowest direction to 11/16-inch, the lower side to be about 1/16-inch from the stockbolt hole at B, the opening to follow the original direction of the lightening holes.

Mark off a width of 11/16-inch at C for the trap spring clearance, taking care to keep it central from the sides of the butt, and with the keyhole saw cut down until the profile at D is met, at the same time inclining the saw downwards towards A, cutting that end about 3/32-inch deeper than at D, then with the "chisel, firmer, ½-inch," cut away the wood between the saw slits.

With the ¼-inch gauge cut the necessary clearance at B and F for the insertion of the pull-through weight, and clearance for head of trap spring screw respectively.

Assemble the Plate, butt, R.S., M.L.E., Mark I*, to butt, raise trap and see that the spring has the necessary clearance, and test weight of trap spring 2 to 3 lb., then insert and remove an oil bottle and pull-through weight to see if they have sufficient clearance.

14363—Rifles, short, M.L.E.—   8 Oct 1908
    Mark III   C
    Converted Mark IV   C

Heavier extractor spring.
LoC 14236 cancelled.

Rifles of the above-mentioned patterns will, in future manufacture, be fitted with an extractor spring, weighing "from 7 to 9 lb." instead of "from 4½ to 5½ lb."

The heavier extractor spring will be issued as spare for all rifles fitted with the bridge charger guide, and may be issued as spare for all other patterns of magazine carbines and rifles.

For purpose of identification this spring will be marked with a figure 3 inside the long end.

14368—Pouches, ammunition, buff, rifle—   13 Oct 1908
    Converted from V.E. pattern 1888, buff—
    20 rounds in chargers—
    Pouches, ammunition, .303-inch, 40 rounds;
      50 rounds, Marks I, II and III; and 50 rounds,
        pattern 1894, Mark I.

    Modification of nomenclature.

    1. With a view to utilizing buff pouches for carriage of .303-inch charger-packed ammunition (LoC 11753), for Royal Garrison Artillery, Army Service Corps, and Army Ordnance Corps, the pouches, as shown in detail column above, have been modified under special instructions issued to the several Commands.

    Such pouches, when altered, will be designated as now shown.

    2. The undermentioned pouches are suitable, and will be utilized for charger-packed ammunition without alteration—
    V. E. pattern 1882, buff—
      Pouches, ammunition—
        Mark V   40 rounds.
        Garrison Artillery   Also A.S.C. and A.O.C.; 40 rounds, &c

    3. That part of the first paragraph of LoC 13713, regarding designation of the altered "Pouch, ammunition," is hereby cancelled.

14424—Grenade, hand (Mark I)   L   6 Jul 1908
    Grenade, hand, for instruction (Mark I)   L
    Dummy.
    Grenade, hand, practice (Mark I)   L
    Dummy; with 2 spare handles.

    Detonator, grenade, hand (Mark I)   L
    Copper.
    Detonator, grenade, hand, for instruction.
      (Mark I)   L
    Copper; dummy.

    Drawings (R.L. 14751 E (1), 15825 (1) and 15831 (1)) of the above-mentioned hand grenades and detonators have been sealed to govern manufacture for Land Service.

    The grenade consists of a brass body with hook for attaching to a waist belt, a cap with needle and safety pin, a partially segmented cast-iron ring, and cane handle with wood block and silk braid tail, detonator and bursting charge.

    The grenade for instruction is similar to the Service one, but has a wood block in place of the charge, and a dummy detonator.

The dummy detonator for instruction is charged with clay and is stamped "Dummy". The instructional grenade has the word "Dummy" painted on it in red, and both the grenade and detonator have holes, .5 inch diameter in the case of the grenade and .15 inch in the detonator, bored through them.

The practice grenade has a beech wood body fitted with a fixed wrought-iron cap, and has no safety pin. It is fitted with a partially segmented iron ring, and weighted with lead to the weight of the Service grenade. Two spare cane handles are issued with each of these grenades.

14427—**Chest, rifle, short, M.L.E. (Mark I)**   L   25 Aug 1908
     Cases—   13 Oct 1908
      10 rifles, short, M.L.E. (Mark I)   L
      6 rifles, short, M.L.E. (Mark I)   L
      2 rifles, short, M.L.E. (Mark I)   L

Modified to suit pattern 1907 sword-bayonet and scabbard.

Chests and cases of the above-mentioned description (LoC 11808, 13105) will, in future manufacture, be fitted internally to carry the pattern 1907 of sword-bayonets and scabbards, (LoC 14170).

Existing chests and cases will be altered locally as may be found necessary, by shifting the cleats on the sides of the chests and cases to the required positions, the existing fittings being used. When packing, a strip of blanket will be wrapped round the hilts of the sword-bayonets to prevent movement.

These alterations will not involve an advance of numeral.

14428—**Scabbard, sword-bayonet, pattern 1907**   3 Dec 1908
       (Mark I)   C
       Brown leather.

Adoption by Navy.

It having been decided that the Navy shall adopt the above-mentioned scabbard (LoC 14170), the Service letter has been altered to "C".

14437—**Boxes**   19 Nov 1908
       Cases
       Chests

Handle cleats to be secured by rivets and screws.

With reference to LoC 13343— The handle cleats of all boxes, cases and chests will, in future manufacture, be secured by rivets and screws instead of screws only.

Existing boxes, cases and chests, when passing through Ordnance Factories or local Ordnance Workshops, for repair, will have the cleats secured by rivets and screws, if it is found necessary to renew the cleats.

14469—Gauges, armourers', pistol, Webley—   18 Nov 1908
    Gauge, distance of cylinder from face
    of body, .052-inch, rejecting   C

Approval having been given to increase the toleration for the distance between the cylinder and face of body in Webley pistols in the hands of troops by .003 inch, the above-mentioned gauge has been introduced.

Officers commanding units and Ordnance officers in charge of depots with armourers attached will demand gauges of the new pattern and return the .049-inch gauges (LoC 9832 and 10160) to store.

For Naval Service— Arrangements will be made by Admiralty for supply of new pattern gauges.

14470—Musket, fencing, short (Mark II)   C   6 Nov 1908
    Converted from rifles, M.H. With spring   3 Sep 1908
    bayonet, but without india-rubber pad.   9 Jan 1909

With reference to LoC 13991— Muskets of the above-mentioned pattern will in future be fitted with a strengthened bayonet, and with a lower band, screw and pin 7 inches from the front of the body.

1. In future manufacture the portion of the bayonet screwed for receiving the cap will be partly screwed and partly plain, with a radius in the corner to prevent breakage at that point. The cap will be correspondingly altered to suit the bayonet.

For purpose of identification these bayonets and caps will be known as No. 2, and will be marked with the figure "2".

2. Muskets in store and in the hands of troops will be fitted locally with the lower band, pin and screw by Army Ordnance Department or regimental armourers in the following manner—

Remove the nosecap screw and pin stud barrel; drive the fore-end with drift fore-end until clear of the body, and then pull the fore-end back clear of the nosecap. Open the band wide enough to pass over the barrel and then close it to its original shape. Replace fore-end and barrel stud pin. Mark off the position of the band on fore-end 7 inches from the front of body, and with the "saw, slitting 7 1/8-inch" cut round the fore-end to a depth of about 1/8-inch, and with chisel and files remove the wood and fit the lower band (a few light taps with a wood mallet will make the band fit the barrel); then screw home the band screw. With an "awl, blade, armourers" of a suitable size bore a hole for the lower band stop pin, and insert the pin.

Ordnance Officers and Officers Commanding will demand the following components for fitting to each musket—
Band, lower, M.H.
Pin, stop, band, lower, M.H.
Screw, band, lower, M.H.

For Naval Service the bands will be fitted to the muskets in accordance with directions that have been issued by the Admiralty.

14522—Box, cartridge, aiming tube, R.F. (Mark I)   L   29 Nov 1904
Wood.

1. New pattern. LoC 13642 amended.

Box, cartridge, aiming tube, C.F. (Mark I)   L
Wood, with tin lining.

2. Nomenclature.

1. A pattern of the box used for packing "Cartridges, aiming tube, R.F." (LoC 13642) has been sealed to govern manufacture.

The top, bottom and sides of the box are made of ¾-inch white deal, and the ends of 7/8-inch elm. The top and bottom are screwed by 8 and 14 brass screws respectively, while the sides and ends are dovetailed. The ends are fitted with 7/8-inch elm cleats and rope handles. Wooden partitions are placed in the box so that movement of the tins during transit is avoided.

Dimensions (exterior for stowage)
Length . . . . .18.0 ins.
Width . . . . .13.8 ins.
Depth . . . . 7.2 ins.

As none of the boxes described in LoC 13642 have been made, the description of the box in that paragraph will be cancelled.

2. The nomenclature of the box described in LoC 10843 has been amended to read as shown above, in order to agree with the nomenclature of the "Cartridges, aiming tube, C.F." (LoC 13642), with which it is packed.

14530—Frogs, buff, bayonet, G.S., Mark II   11 Feb 1909

    Modification to suit sword-bayonet and scabbard, Pattern 1907.

    If difficulty be experienced by units, on receiving the sword-bayonet, Pattern 1907 (LoC 14170), in fitting them into frogs of the above description, the following alteration may be made thereto—

        Remove the strap and buckle parts, cut away ¾ inch of the front of the frog at the top, and slightly trim the slot for the scabbard button, rounding outwards; then cut away sufficient of the folded-back part to allow the strap and buckle parts, when replaced, to lay between the part cut away and the front of the frog. Rivet the front at each top corner; also the strap and buckle parts after re-sewing.

    A sample frog will be issued on indent to guide alterations.

14575—Haversacks—   1 Feb 1909
    Green rifles
    Khaki, dyed
    White canvas

    Obsolete.

    Haversacks of the above-mentioned patterns are declared obsolete. Any now in possession of troops will be retained until worn out, or until replaced under special instructions. All such haversacks in possession of the Army Ordnance Department, or returned to store from time to time by units, will be disposed of under instructions which have been issued to Chief Ordnance Officers by the Deputy Director of Ordnance Stores, Woolwich Arsenal.

    A similar course will be taken with regard to any other haversacks of old pattern, with the exception of the undermentioned—
    Haversacks—
        Drab.
        Drab, N.P. (D.S.)
        Drab, N.P. (M.S.)
        G.S.,   which will be retained for use.

14606—Case, 20 barrels, M.L.M. carbine, pattern A   L   29 Jan 1909
    Case, 200 bayonets, M.H., long (Mark I)   C   25 Feb 1909
    Case, regimental, 10 carbines, M.L.M.,
        cavalry, pattern A   L
    Chest, carbine—
        Without fittings, D   L
        With fittings, D   L

Obsolete.

Cases and chests of the above-mentioned description (LoC 6372, 7952, 10797), both "Plain" and "Prepared for foreign service" are hereby declared obsolete.

The cases and chests will be used up as ordinary "Cases, wood, packing."

14639–Cut-off, M.L.E. rifle, short, Mark III   C   5 Mar 1909
 All rifles, short, M.L.E.; also rifles,
 charger-loading, M.L.M. and M.L.E.

A pattern of the above-mentioned cut-off has been approved to govern future manufacture. It differs from that described in LoC 13742 in the following particulars—

The outer and inner stop nibs on the rear end have been lengthened; the inner stop nib altered in angle to strengthen it; and the outer stop nib turned up to take a bearing on the solid portion of the body above the cut-off slot.

Rifles in the service in which the stop nib of the old pattern cut-off has cut through the thin portion of the body at the rear end of the cut-off slot, should be made serviceable by demanding and fitting the new pattern cut-off.

When fitting, armourers may have to slightly increase the clearance in the magazine case at A, on account of the lengthened inner stop nib of the new pattern cut-off.

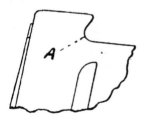

14651–Cover, rifle, Mark II   18 Dec. 1908
 Canvas, with strap and buckle.   22 Apr 1909

A pattern of the above-mentioned article has been sealed to govern future manufacture, and will replace the covers referred to in LoC 12774.

The cover is made of canvas (cutch dyed), the inner flap is bound with tape, and the outer with leather; it is also fitted at the mouth with a leather strap, chape and buckle.

14673—Box, detonators, grenade, hand (Mark I)   L   5 Feb 1909
    Tin, to hold 10.

A pattern (design R.L. 16098) of the above-mentioned box has been sealed to govern manufacture for Land Service.

The box is of tin, painted red externally, lined with asbestos, and fitted with a cork block bored to receive 10 detonators, hand grenade (LoC 14424). A sheet of cork, fitted with a loop of tape to facilitate removal, is placed over the detonators and the box closed by a soldered tin band. Five boxes are packed in a "Case, wood, packing."

14674—Box, ammunition, S.A., 600 rounds,   19 Mar 1909
    .303-inch, in chargers (Mark I)   31 Mar 1909
    Mahogany, with tin lining; or 840, or 800
    rounds, in paper packets.

Packing with 800 rounds of packet ammunition.
Nomenclature.

The above-mentioned box (LoC 12117), when packed with ammunition in paper packets for West African Frontier Force, will in future contain 800 instead of 840 rounds, as the latter amount makes the package too heavy.

The details apended to the nomenclature of the box will accordingly be amended as shown above. The use of a packing piece is not entailed by the new packing.

14676—Mandril, scabbard, sword-bayonet,   26 Apr 1909
    pattern 1907, blocking leather (Mark I)   C
    And Mark II

A pattern of the above-mentioned mandril, for use of armourers, has been sealed to govern manufacture.

14677—Scabbard, sword-bayonet, pattern 1907   23 Apr 1909
    (Mark I)   L
    Brown leather.

LoC 14428 cancelled.

In consequence of the introduction of the Mark II scabbard (LoC 14678), no Mark I scabbards will be supplied to the Navy and LoC 15528 is cancelled accordingly.

14678—Scabbard, sword-bayonet, pattern 1907 11 Dec 1908
    (Mark II)   C
    Brown leather.

A pattern of the above-mentioned scabbard has been approved for future manufacture. It differs from the Mark I scabbard described in LoC 1470 only in being fitted with an external instead of internal chape.

The following table shows the interchangeability of the components—

Component	Remarks
Chape	Special
Lace, iron, short (3)	Common (2 for locket, 1 for chape)
Lace, iron, long (1)	Special
Leather, finished	Special
Locket	Same as pattern 1907, Mark I
Rivet, spring locket (6)	Same as pattern 1888
Spring, locket (2)	Same as pattern 1907, Mark I

Scale $\frac{1}{5}$

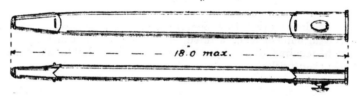

14687—Web equipment, pattern 1908— 4 May 1909
    Belt, waist—
        Large
        48 inches.
        Medium
        44 inches.
        Small
        40 inches.

Addition of medium size.

With reference to LoC 14288, the waist belt will in future be provided in three sizes, the overall lengths being 48, 44 and 40 inches respectively.

14710—Carbines 4 May 1909
    M.L.M.
    M.L.E.

Rifles—
M.L.M.
M.L.E.
Short, M.L.E.
M.L.M., C.L.
M.L.E., C.L.
Short, .22-inch, R.F.
A.T.

Rounding of the edge on the front face of bolt-head.

In future manufacture, or conversion, bolt-heads for the above-mentioned arms will have the edge on the front face rounded.

After fitting and adjusting spare bolt-heads in accordance with instructions given in LoC 12611 and 13611, armourers will round the edge to a radius not exceeding .02-inch.

14711—Chest, rifle, with fittings, "C"   C   31 Dec 1908
   Wood, 20 arms with side-arms; all M.L.M.,   27 May 1909
   M.L.E., M.H. and M.M.

To take either long or short rifles.

A drawing (R.C.D. 11848A) has been approved as a record of a certain number manufactured, and to govern the conversion, as may be required, of the chests mentioned in LoC 7269 to take either long or short rifles.

The conversion, which consists in providing removeable false ends, additional butt slips, and means for adjusting the bridge fittings, will be carried out as follows—
For Naval Service— The chests at present in the Service will be converted locally to take the short rifle for transport at home.
For Land Service— The chests will be converted locally in accordance with the accompanying drawing, the necessary material being obtained locally.

Printed instructions for the arrangement of the fittings for the long or short rifle will be supplied to officers concerned on demand, as required, and affixed to the inside of the chest.

The external dimensions and tonnage remain unaltered.

14712—Mandril, scabbard, sword, Cavalry,   30 Jun 1909
   pattern 1908 (Mark I)   L
   Steel.

A pattern of the above-mentioned mandril for use of armourers has been sealed to govern manufacture.

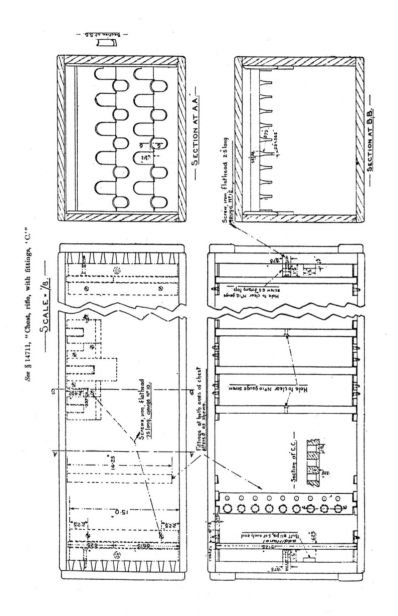

14755—Mandril, scabbard, sword-bayonet,   20 Jul 1909
    pattern 1907, Chape. (Mark I)   C
    Steel.

    1. New pattern.

    Mandril, scabbard, sword-bayonet,
    pattern 1888, Locket. (Mark I)   C
    Steel; and II* and Naval Mark I; also
    patterns 1903 Land and Naval, and pattern 1907.

    2. Extended use.

    1. A pattern of the above-mentioned mandril, for use of armourers in fixing the laces which hold the chape to the scabbard, has been sealed to govern manufacture.

    2. In consequence of this mandril (LoC 6379) being suitable for use in fixing the laces which hold the locket to the scabbard in other patterns, its use has been extended as shown above.

14756—Musket, fencing (Mark V)   L   25 Jun 1909
    With spring bayonet and indiarubber pad.

    1. Altered nomenclature.
    2. Extended use.

    1. The nomenclature of the "Musket, Mark V" described in LoC 8369, will, in future, be as shown above.

    2. In consequence of the length of this musket being about the same as the length of "Rifles, short, M.L.E.," with pattern 1907 sword-bayonet fixed, it will be issued for use of troops armed with the before-mentioned rifle and sword-bayonet until the stock is used up, when the "Musket, fencing, Mark VII" (LoC 14757) will be taken into use.

14757—Musket, fencing (Mark VII)   L   26 Mar 1909
    Converted from Rifle, M.H., with spring   20 Apr 1909
    bayonet, but without indiarubber pad.

    A pattern of the above-mentioned musket has been approved to govern conversion as may be ordered.

    It differs from the "Musket, fencing, short, Mark III" described in LoC 13991 & 14470, in the following particulars—

    The barrel and fore-end are left 3 3/8 inches longer to make the musket equal in length to the short rifle with the pattern 1907 sword-bayonet fixed.

The nosecap is screwed on the barrel and fixed by a pin passing through the nosecap and the underside of the barrel.

Total length of musket (over all) . . . . .5 feet 1½ inch.
Total weight of musket. . . . . .8 lb. 14 oz.
Balance from butt end . . . . .24 inches.

14758—Rifle, charger-loading, M.L.M. (Mark II)   1 Feb 1909
                                                                                            6 Aug 1909
Change in nomenclature of certain components (LoC 13992).

It having been decided that no more rifles are to be converted to the above-mentioned pattern, and that those already converted are to be altered to Rifles, C.L.M.L.E., Mark I*, by fitting Bolts, breech, M.L.E.C.L.R., I*, assembled, the components shown as "M.L.M.C.L.R., II" will now be shown as "M.L.E.C.L.R., I*" with the following exceptions, which will retain their present nomenclature—

Barrels, M.L.M. C.L.R., II Blades, foresight, M.L.M.C.L.R., II Protectors, foresight, M.L.M.C.L.R., II	Special to Rifles, C.L.M.L.E., I*, having barrels with "Metford" rifling which will be replaced by barrels with "Enfield" rifling as they become worn out.
Bolts, breech, M.L.M.C.L.R., II	Now obsolete.

Rifles altered in accordance with the foregoing will be the same in all detail as Rifles, C.L.M.L.E., Mark I*, converted from Rifles, M.L.M., Mark II*, vide LoC 13992.

14795—Web equipment, pattern 1908—   17 May 1909
    Carriers, intrenching tool—   24 Aug 1909
    Head   8 Oct 1909
    Helve

Patterns of the above-mentioned articles have been sealed to govern supply for implements, intrenching, pattern 1908.

The carrier for the head of the tool is bag shaped, is made of the same material as the "pack" and fitted with a short web closing strap with snap fastener; it is also fitted with two chapes and buckles for attaching the carrier to the front end of the right brace, and to the tab on the waist belt on that side. In inserting the tool it should be put into the carrier pick end first and as horizontally as possible, to prevent the sharp edge being caught by the bottom of the bag.

The helve carrier is made in the form of two double loops of webbing connected by a chape, fitted with a buckle for fastening the

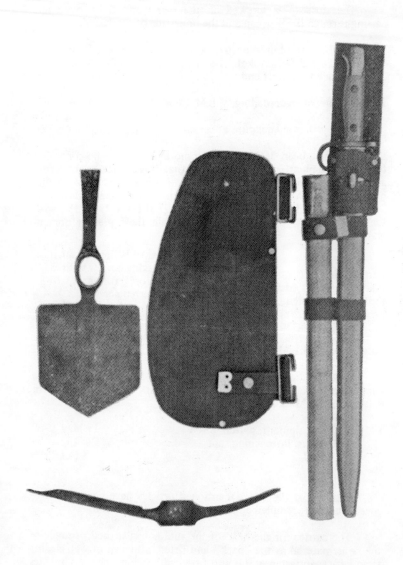

200

carrier to the tab on the back of the frog. One set of loops is passed over the bayonet scabbard, the helve is passed through the other, thus securely connecting the helve and bayonet scabbard together. The part of the top loop which passes round the helve is fitted with two brass eyeleted tips and a fastener. The helve is carried in front of the scabbard.

14796—Implement, intrenching, pattern 1908—   22 Oct 1908
   Head   23 Jul 1909
   Steel, shovel and pick ends.   8 Oct 1909
   Helve
   With ferrule

   New patterns.

A pattern of the above implement has been approved to govern supply. See illustration published with LoC 14795.

The head, which is made of special steel, has at one end a pointed shovel-shaped blade, which is strengthened in the centre by a mitre or tapered rib extending from the eye to the point of the blade; the other end is shaped similarly to the flat or chisel end of a pickaxe.

The helves are made of either ash or hickory, and are oval in cross section throughout, they are 16½ inches long, with a ferrule on the largest end.

The combined weight of head and helve is approximately 1 lb. 3½ oz.

14830—Cartridges, S.A., dummy drill, .303-inch   18 Oct 1909
   rifles or carbines (Marks I and II).   L

   Obsolete.

The above-mentioned cartridges (LoC 6057, 9519) are hereby declared obsolete. Existing stocks of these Marks will be brought to produce locally.

14835—Slide, backsight, M.L.E. C.L. rifle,   23 Jun 1909
   Mark I*, No. 2   L
   Assembled.

A pattern of the above-mentioned slide has been approved to govern future manufacture.

The following is a list of its components, showing which are suitable for, or common to, the No. 1 slide (LoC 13992). Those which are of new pattern are shown as No. 2—

Component	Remarks
Head, screw, windgauge, No. 2	Not issued separately.
Nut, clamping, slide, backsight, No. 2	Suitable for No. 1 slide if issued with No. 2 windgauge screw.
Pin, fixing, head, screw, windgauge, No. 2	
Screw, windgauge, backsight, No. 2 (assembled with head and fixing pin).	Suitable for No. 1 slide if issued with No. 2 clamping nut.
Slide, backsight, No. 2 (unassembled)	
Spring, slide, backsight	Common.
Spring, windgauge, backsight	Common.
Stud, clamping, slide, backsight	Common.
Windgauge, backsight	Common.

14866–Scabbards, sword-bayonet—   12 Oct 1909
    Pattern 1888–
    Land, Marks II and II*   L

    Pattern 1903–
    Land, Mark I   L
    Pattern 1907, Mark I   L

Fitting with external chape on repair.

With reference to LoC 11151, 11811, 12612 and 14170– Scabbards of the above-mentioned patterns, referred to therein, when undergoing repairs of a nature requiring the fitting of a new chape or leather, will be fitted with an external chape of the pattern for scabbard, sword-bayonet, pattern 1907, Mark II, referred to in LoC 14678.

14899–Box, cartridge, aiming tube, R.F. (Mark I)   L   28 Oct 1909
    Wood, with 10 tin boxes.

    Nomenclature.

In future the 10 tin boxes in which ammunition is packed will be regarded as components of the "Box, cartridge, aiming tube, R.F." (LoC 14522). The detail of nomenclature has in consequence been amended to read as shown above.

14902–Musket, fencing (Mark VII)   L   26 Mar 1909
    Converted from Rifle, M.H.; with spring   25 Apr 1909
    bayonet, but without india-rubber pad.

    New pattern.
    LoC 14757 cancelled.

A pattern of the above-mentioned musket has been approved to govern conversion as may be ordered.

It differs from the "Musket, fencing, short, Mark II" described in LoC 13991 and 14470 in the following particulars—

> The barrel and fore-end are left 3 3/8 inches longer to make the musket equal in length to the short rifle with the pattern 1907 bayonet fixed.
>
> The nosecap is screwed on the barrel and fixed by a screw passing through the nosecap and bayonet bush into a recess in the top side of the barrel. The metal of the nosecap is expanded into the slot of the screw by a centre punch to prevent the screw working loose.
>
> The chamber of the barrel is fitted with a wooden plug in two parts to enable the bayonet spring of the "Musket, fencing, short, Mark II" to be used.
>
> Total length of musket (overall).........5 ft. 1½ ins.
> Total weight of musket...............9 lb. 5 oz.
> Balance from butt end..............24 ins.

The following is a list of the components, showing which are special to this musket, and which are common to other arms. Those which are common to other arms are denoted by the word "Common" and their interchangeability is as stated in the "Priced Vocabulary of Stores" &c.

LoC 14757 is hereby cancelled.

Nomenclature	Remarks
Balance, weights, musket, Mark VII	Special.
Bands, lower, musket, short, Mark II	Common.
Barrels, musket, Mark VII	Special. With body (with wood blocks and fixing pegs); nosecap and screw; bush, stop bayonet and fixing screw, and plugs, barrel.
Bayonets, musket, short, Mark II, No. 2	Common. With cap and fixing pin.
Blocks, wood, body, musket, short, Mk II, No. 2	Common. Not issued separately.
Bodies, musket, short, Mark II.	Common. With blocks, wood (in 2 parts) and 4 fixing pegs. Not issued separately.
Bolts, stock, musket, short, Mark II	Common.

Component	Remarks
Bushes, stop, bayonet, musket, short, Mark II, No. 2	Common, not issued separately.
Caps, bayonet, musket, short, Mark II, No. 2	Common, not issued separately.
Caps, nose, musket, Mark VII	Special, not issued separately.
Pins, musket, fixing, cap, bayonet, short, Mark II.	Common.
Pins, musket, stop, band, lower, short, Mark II	Common.
Pins, musket, stud, barrel, short, Mark II	Common.
Plates, butt, musket, Mark V	Common.
Plugs, barrel, musket, Mark VII (2)	Special.
Screws, musket, band, lower, short, Mark II	Common.
Screws, musket, butt plate, Mark V (2)	Common.
Screws, musket, cap, nose, short, Mark II	Common.
Screws, musket, fixing, bush, stop, bayonet, Mark VII.	Special.
Springs, bayonet, musket, short, Mark II	Common.
Stocks, musket, butt, short, Mark II	Common.
Stocks, musket, fore-end, Mark VII	Special.
Stops, stud, musket, barrel, short, Mark II	Common.
Washers, bolt, stock, musket, short, Mark II	Common.

14903—Muskets, fencing—   26 Oct 1909
  With spring bayonet.
    Mark IV   L
    With india rubber pad.
    Mark V   L
    With india rubber pad.

    1. Altered nomenclature.

    Muskets—
    Fencing—
    With spring bayonet.
      Mark IV   L
      With india rubber pad.
      Mark V   L
      With india rubber pad.
      Mark VI   L
      Wood; with india rubber ball.
    Short—
      Mark I   L
      With india rubber pad.
      Mark II   C

Converted from rifles, M.H.;
without india rubber pad.

2. To be used up.

1. The nomenclature of "Muskets, Marks IV and V," described in LoC 7406 and 8369, will in future be shown as above.

2. Muskets, fencing, of the above-mentioned pattern will be used up before taking into use the Mark VII pattern described in LoC 14902.

LoC 14756 is hereby cancelled.

14904—Rifle, Short, M.L.E.   10 Nov 1909
    All Marks.

Omission of "pin, stop, band, outer."

In future manufacture, the "Pin, stop, band, outer" will not be fitted to new stocks, fore-end, for the above-mentioned rifles, as the outer band is retained by the recess in the stock, fore-end, and hand-guard.

14905—Screws—   9 Dec 1909
    Band, lower, M.L.M. rifle, Mark II.
    Swivel, piling, M.L.M. rifle, Mark II.

In future manufacture, spare screws of the above description will have a longitudinal hole bored at the screwed end to enable the end of the screw, when screwed home, to be expanded with a centre punch, thus preventing the movement of the swivels working the swivel screws loose.

14936—Rifle, short, magazine, Lee-Enfield   4 Jan 1908
    (Mark I**)   N   25 Jun 1908
    Converted from Rifle, short, M.L.E.,   5 Jul 1909
    Mark I.   22 Oct 1909

    Rifle, short, magazine, Lee-Enfield,
    converted (Mark II**)   N
    Converted from Rifle, short, M.L.E.,
    converted, Mark II.

    Rifle, short, magazine, Lee-Enfield,
    converted (Mark II***)   N
    Converted from Rifle, short, M.L.E.,
    converted, Mark II*

Patterns of the above-mentioned rifles have been sealed to record the above-mentioned conversion.

The conversion (which is carried out in Naval Ordnance depots, Chatham, Portsmouth, and Plymouth, in accordance with Admiralty letters of 4th January, 1908, G. 16202/08/247-9, and of 1st July, 1908, G. 8313/08/13378-80) consists in fitting a "Blade, foresight, M.L.E., rifle, short, Mark III" (LoC 13853) and a higher windgauge, with top edge hardened, and with a U-notch, to the backsight to correspond.

The windgauge is special to the above-mentioned rifles.

In addition to the above-mentioned conversion the following additions and alterations have been made—

LoC 13313. Body— Sharp edges on charger guide on left side and on charger guide stop on right side have been removed to prevent damage to clothing and equipment.

LoC 13509. Backsight— The rear end of the sight leaf is bored to receive two spiral springs to tension the fine adjustment, and a new screw, fine-adjustment, having a larger head and finer milling, is fitted, having a tension spring under the head to prevent the screw jarring loose in firing the rifle.

Guard, hand, rear— The rear handguard is fitted with a double spring, fixed by two longer rivets placed in a central position in the length of the spring.

Guard, trigger— The trigger guard is provided with a lug at the front to carry a sling swivel and screw when required in that position and the magazine link loop is removed.

Link, magazine— The magazine link has been removed.

Magazine— The magazine link loop has been removed.

Spring, platform, magazine— A retaining nib has been riveted to the spring to prevent any movement of the platform on the spring.

Stock, fore-end— The stock, fore-end, has been recessed to receive a stud and spring to centre the barrel in the nosecap. It is also suitably recessed for the double spring of the handguard.

Plate, butt— The "Plate, butt, M.L.E. rifle, short, Mark I*, No. 2" (LoC 13742), has been fitted, and a wad, stockbolt, is inserted in the stock, butt, except in the case of the converted Mk II***, which was already fitted with the M.L.M. rifle, Mark II, buttplate and a wad inserted before conversion from Mark II*.

Stock, butt— The stock, butt, has been suitably recessed to receive the butt plate, except in the case of the converted Mark II*** which was already fitted as described above.

LoC 13511— Screw, keeper, striker— A keeper screw (which can be turned by a coin), for retaining the striker in the cocking piece, has been fitted in place of the nut, screw, and spring, keeper, striker, except in the case of the converted Mark II***, which was already fitted before conversion from Mark II*.

LoC 13549— Bolt, locking— The sharp edges of the locking bolt have been removed to prevent damage to clothing and equipment, except in the case of the Mark II***, which was already so altered before conversion from Mark II*.

LoC 13577— Screws, swivel— To prevent the swivel screws working loose, they are bored longitudinally at the screwed end to enable the end of the screw, when screwed home, to be expanded by the armourer, with the centre punch issued as part of the foresight adjusting tools, except in the case of the converted Mark II***, which was already so altered before conversion from Mark II*.

LoC 13742— Guide, charger, No. 2 and screw— The No. 2 charger guide and screw have been fitted.

LoC 14070— Nosecap and stock, fore-end— The nosecap and stock, fore-end, have been numbered to correspond with the barrel number.

Head, breech, bolt— The edge on the front face of the bolthead has been rounded in accordance with the instructions given in LoC 14710 (vide Admiralty letters of 6th July, 1909, G. 9816/09/12707-9, and of 23rd August, 1909, G. 9816/09/15777-9).

All spare parts in store at Naval Ordnance depots are being altered as affected by the above-mentioned modifications. When the conversion described above has been carried out, the marking on the body and butt of rifles will be amended accordingly (vide Admiralty letter, dated 22nd October, 1909, G. 15883/09/19888-90).

14937—Rifles, short, M.L.E.—   4 Nov 1909
    (Mark III)
    Converted (Mark IV)

Fitting washers to take up backlash of windgauge screw.

In cases where backlash of windgauge screw exists owing to the distance between the removeable head of the windgauge screw and the right side of the windgauge being too great, one or more washers .005 inch thick will be placed in the recess on the left side of the windgauge under the small slotted head of the windgauge screw.

Before fitting washers, armourers should be careful to see that the backlash is not due to any defect in the windgauge screw spring, which, if defective, should be exchanged. Officers commanding units will put forward demands for these washers as required.

The washers will be described as "Washers, screw, windgauge, M.L.E.R.S., Mark III" and for Naval Service, will be supplied in boxes of 50.

14938–Rifles, short, M.L.E., converted (Mark IV)   22 Oct 1909

Rounding of the edge on the front face of bolt heads in rifles in Naval Service.

Rifles of the above-mentioned pattern already issued to Naval Service will have the edge on the front face of boltheads rounded in accordance with the instructions given in LoC 14710. Spare parts in store at Naval Ordnance depots at Chatham, Portsmouth and Plymouth, in accordance with Admiralty letters of 6th July, 1909, G 9816/09/12707-9, and of 23rd August, 1909, G 9816/09/15777-9.

14945–Valise, equipment, pattern 1888, buff–   22 Dec 1909
Belts, waist, G.S., Marks I to III.

Modification of, for R.A.M.C.

It has been decided that when the stock of V.E. pattern 1882 belts, waist, G.S., has been used up, belts of patterns described above will be modified for issue to units of R.A.M.C. as required.

The modification consists in the removal of the tongues from rear buckles, and also of the 1-inch brass loops from the runners at each end of the belt. Belts of 1888 pattern at outstations when required for R.A.M.C. will be altered prior to issue, and described as-

V.E. pattern 1882, buff–
Belts, waist, G.S.

14971–Guides, muzzle–   4 Nov 1909
    No. 1 Mark II   C
    For rifles, M.L.M., M.L.E., (including charger-loading), and M.E., and carbines, Artillery, M.M. and M.E.
    No. 2, Mark II   C
    For rifles, short, M.L.E., and carbines, Cavalry, .303-inch.

1. New patterns.

   Guides, muzzle—
     No. 1 Mark I   C
     For rifles, M.L.M., M.L.E., (including charger-loading), and M.E., and carbines Artillery, M.M. and M.E.
     No. 2, Mark I   C
     For rifles, short, M.L.E., and carbines, Cavalry, .303-inch.

2. Change in nomenclature.

   1. Patterns of the above-mentioned tools have been sealed to govern future manufacture. They differ from the Mark I patterns described in LoC 11079 and 11851 in the length of the cleaning rod hole, which has been increased from ¼ inch to ¾ inch, and in being marked with the figures 1 and 2 respectively to denote pattern.

   2. In consequence of the introduction of the new patterns the nomenclature of the muzzle guides described in LoC 11079 and 11851 has been amended as shown above.

   3. In Naval Service the existing stock of Mark I patterns will be used up.

14972—Swords, practice, Mounted Corps   L   18 Jan 1910
   Converted from cavalry swords.

   Obsolete.

   Swords of the above-mentioned type are hereby declared obsolete and will be returned to Army Ordnance Department for disposal.

15000—Carbines—   1st Mar 1910
   M.E.
   M.L.E.
   M.M.

   Rifles—
   M.E.
   M.L.E.
   M.L.E., charger-loading
   M.L.M.

Letter "E" on knoxform of barrels to be omitted.

With reference to paragraphs 9124, 11078 and 11498— In future manufacture of spare barrels for the above-mentioned arms, the letter "E", denoting that the barrels are rifles with "Enfield" rifling, will not be stamped on the knoxform. In the case of barrels for "Rifles, M.L.M., Mark I*" the pattern numeral "I*" also will be omitted.

15001—Sword-bayonet, pattern 1907, Mark I   C   16 Mar 1910

Sharpening before troops proceed on active service.

The edge of sword-bayonets of the above-mentioned pattern will be sharpened before troops proceed on active service, in accordance with the instructions given in LoC 9206. The sharpening will be restricted to that portion of the blade which has been already reduced to .01 inch thick.

For Naval Service, the sharpening for active service will be carried out in accordance with instructions issued by the Admiralty.

15002—Tube, aiming, .23-inch, M.L.E. rifle,   4 Feb 1910
  short (Mark I)   N
  Morris; including extractor and screw, set
  nut and washers.

A pattern of the above-mentioned tube has been sealed to govern manufacture.

It differs from the "Tube, aiming, Lee-Metford magazine rifle, .303-inch," described in LoC 6602, being about 5 inches shorter to suit the shorter barrel of the "Rifle, short, M.L.E."

The instructions for use given in LoC 6602 will apply to this tube, with the exception of that part of the last paragraph referring to the cleaning rod.

Length of tube overall . . . . .26 inches.

INDEX

		LoC No.
Accoutrements, naval, P' 1901	Introduction	11110
Accoutrements, naval, P' 1901	New pattern	11659
Accoutrements, naval, P' 1901	Nomenclature	11111
Accoutrements, naval, P' 1901	Pistol case	11202
Accoutrements, naval, P' 1901	Pockets, ctge.	12697
Accoutrements, various	New patterns	12423
Accoutrements, various	Obsolete	12423
Action, skeleton, M.L.E., short (Mk III)	Introduction	14207
Action, skeleton, pistol, Webley	Introduction	10392
Action, skeleton, rifle, short, M.L.E. (Mk I)	Introduction	12220
Action, skeleton, rifle, short, M.L.E., cond. (Mk I)	Introduction	12221
Aim corrector (Mk II)	Introduction	12421
Bag, armourers (Mk I)	Introduction	13679
Bandolier equipment, P' 1903	Carrier	12471
Bandolier equipment, P' 1903	Introduction	12389
Bandolier equipment, P' 1903	Cover	12526
Bandolier equipment, P' 1903	New patterns	13273
Bandolier, leather, .303-inch ammun.	Nomenclature	10764
Bandolier, web	Introduction	11266
Bandolier, 90 rounds (Mk I)	Introduction	12424
Belt, pouch, buff, Cavalry, Line	Obsolete	10561
Belt, waist, brown, cavalry	New pattern	12425
Belt, waist, brown, Military Mounted Police	Obsolete	12425
Belt, waist, brown, pistol	Obsolete	12425
Belt, waist, brown, Warrant Officers	New pattern	12223
Belts, pouch & waist, various issues	Obsolete	12223
Belts, "Sam Browne" Mk I	Introduction	10440
Belts, "Sam Browne" Mk II	Introduction	11267
Belts, waist, black, warders	Obsolete	10639
Belts, waist, black, warrant officers	LoC cancelled	13754
Belts, waist, buff, cavalry, P' 1885	Alteration	10765
Belts, waist, buff, warrant officers	LoC cancelled	13754
Bottle, oil, Mk III	Introduction	10122
Bottle, oil, Mk IV	Introduction	13470
Box, ammun., small arm, Mk XI	Altered dimens.	13316
Box, ammun., small arm, Mk XIII	Obsolete	10030
Box, ammun., small arm, pistol, Enfield	Alteration	10751
Box, ammun., small arm, 1,000 rounds, .303-in., in chargers	Amendment	14232
Box, ammun., S.A., home & special (Mk XIV)	Brass pin	10505
Box, ammun., S.A., (Mk XV)	Introduction	10749
Box, ammun., S.A., 600 rds., .303-in. chargers	New pattern	12117
Box, ammun., S.A., 600 rds., .303-in. chargers	Packing	14674
Box, ammun., S.A., 750 rds; .303-in (Mk I)	Introduction	10750
Box, cartridge, aiming tube, C.F. (Mk I)	New pattern	14522
Box, cartridge, aiming tube, R.F. (Mk I)	New pattern	14522
Box, cartridge, aiming tube, R.F. (Mk I)	Nomenclature	14899
Box, detonators, grenade, hand (Mk I)	Introduction	14673
Box, rod, tool, clearing	Introduction	13679
Boxes	Alteration	14437
Boxes, ammunition, S.A.	Closing plate	11632
Boxes, ammunition, S.A.	Labelling	10098

Boxes, ammunition, S.A.		Labels	12155
Boxes, ammunition, S.A.		Linings	10844
Boxes, ammunition, S.A.		Lot markings	14204
Boxes, ammunition, S.A.		New labels	11980
Boxes, ammunition, S.A.		Red cross	12073
Boxes, ammunition, S.A.		Red cross	12292
Boxes, ammunition, S.A., Pistol, Enfield		Fastening	12156
Boxes, ammunition, S.A., 780 & 750 rds.		Obsolete	12117
Boxes, ammunition, S.A., 1,000 rds., .303-in.		Dimensions	12157
Boxes, ammunition, S.A., .303-in, half, naval		Fastening	12156
Brush, cleaning, aiming tube		Introduction	13855
Bucket, rifle, Cavalry (Mk I)		Introduction	12714
Bucket, rifle, Cavalry (Mk II)		Introduction	13969
Bucket, rifle (Mk IV)		Alteration	11856
Bucket, rifle (Mk IV)		Alteration	13715
Bucket, rifle (Mk V)		New pattern	11208
Carbine, magazine, Lee-Enfield		for bayonet	10220
Carbine, magazine, Lee-Enfield; P'1888 byt.		Sighting	11807
Carbine, magazine, Lee-Metford		Enfield barrel	11078
Carbines		Alteration	14710
Carbines		Barrel marking	15000
Carbines		Short lead	14234
Carbines		Butt stock	11289
Carbines, Artillery & Cavalry		Enfield barrel	11498
Carbines, Artillery & Cavalry		Nosecap	10687
Carbines, Artillery & Cavalry, Martini		Sight wings	13680
Carbines, .303-inch		Worn barrels	11033
Carbines & rifles		Modification	13136
Carbines & rifles		Spare boltheads	13611
Carrier, water bottle		New pattern	12994
Ctge, aiming tube, C.F. (Mk II)		Nomenclature	13642
Ctge, aiming tube, R.F. (Mk I)		Introduction	13642
Ctge, signal, Very's, white (Mk III)		Introduction	10117
Ctge, S.A., ball, M.H. & Snider		Letter altered	13310
Ctge, S.A., ball, pistol, Webley (Mks II & III)		Obsolete, make	11398
Ctge, S.A., ball, pistol, Webley (Mks II & III)		New mark	10273
Ctge, S.A., ball, .303-in., cordite		Marking	14131
Ctge, S.A., ball, .303-in., cordite		Stamping	14319
Ctge, S.A., ball, .303-in., cordite (Mk II)		Continuation	11397
Ctge, S.A., ball, .303-in., cordite (Mks IV & V)		Obsolete	11397
Ctge, S.A., blank, .303-in., B.P., without bullet (Mk II)		Obsolete	11413
Ctge, S.A., blank, .303-in., cordite, with mock bullet (Mk VI)		Introduction	11317
Ctge, S.A., dummy, drill, .303-in. (Mk I & II)		Onsolete	14830
Ctge, S.A., dummy, drill, .303-in. (Mk III)		Introduction	11832
Ctge, S.A., dummy, drill, .303-in.		Alteration	13926
Ctge, S.A., shot, Snider, special (Mk I)		Alteration	13043
Ctges, signal, Very's		Use extended	12878
Ctges, S.A., ball, M.H.		Introduction	11752
Ctges, S.A., blank, .303-in.		Alteration	13544
Ctges, S.A., blank, .303-in.		Altered use	13737
Ctges, S.A., blank, .303-in.		Nomenclature	13083
Ctges, S.A., dummy, drill, .303-in. (Mks I-III)		LoC cancelled	12597
Ctges, S.A., dummy, drill, .303-in. (Mks I-III)		Modifications	12549
Ctges, S.A., dummy, drill, .303-in. (Mks I-III)		Modifications	12877

Ctges for Instruction, Small Arm	Holes drilled	10094
Ctges for Instruction, Small Arm	Precautions	10746
Case, brown, pistol, "Sam Browne" Mk I	Introduction	10440
Case, brown, pistol, "Sam Browne" Mk II	Introduction	11267
Case, regimental, 10 carbines, M.L.M.	Strengthening	10797
Case, regimental, 10 rifles, M.L.M.	Conversion	12567
Case, regimental, 10 rifles, short, M.L.E.	Conversion	12763
Case, sword-bayonet or scabbard, P' 1903	New pattern	12991
Case, 200 sword-bayonets, P' 1888	Strengthening	11150
Case, 200 sword-bayonets, P' 1903	Introduction	11849
Cases	Alteration	14437
Cases, various	Alteration	14427
Cases, various	New patterns	13105
Cases, various	Obsolete	14606
Cases, 8 & 4 rifles	Obsolete	13266
Charger, .303-in. cartridge (Mk I)	Introduction	11753
Charger, .303-in. cartridge (Mk II)	Introduction	13465
Chest, carbine, with & without fittings D	Obsolete	14606
Chest, carbine, with fittings D	Strengthening	10797
Chest, rifle, short, M.L.E. (Mk I)	Alteration	13056
Chest, rifle, short, M.L.E. (Mk I)	Alteration	14427
Chest, rifle, short, M.L.E. (Mk I)	Introduction	11808
Chest, rifle, with fittings	Introduction	14711
Chests	Alterations	14437
Chests, rifles & carbines	Strengthening	11150
Cordite, T & MDT	Sizes	11975
Cover, cuirass (Mk I)	Undyed canvas	11660
Cover, rifle Mk II	Introduction	14651
Cover, tin, mineral jelly	New pattern	12327
Cuirass	New pattern	12470
Cut-off, rifle, short, M.L.E. (Mk I)	Alteration	12129
Cut-off, rifle, short, M.L.E. (Mk I)	Extended use	13612
Cut-off, rifle, short, M.L.E. (Mk I)	Introduction	11850
Cut-off, rifle, short, M.L.E. (Mk III)	Introduction	14639
Detonator, grenade, hand (Mk I)	Introduction	14424
Dirk (Mk II)	New pattern	11290
Drivers, screw, armourers, medium	Obsolete	11291
Flannelette (Mk III)	Introduction	12603
Frog, buff, sword	Alteration	10258
Frog, sword, saddle, R.A.	Obsolete	11337
Frogs, buff, bayonet, G.S. Mk II	Alteration	14530
Gauge, armourers; magazine rifles & carbines	Alteration	13965
Gauge, armourers', pistol, Webley	Obsolete	10160
Gauge, armourers', pistol, Webley	Alteration	14469
Grenade, hand (Mk I)	Introduction	14424
Guard, hand, front, M.L.M. rifle	New pattern	14069
Guards, hand, M.L.M. & M.E. rifles	Repair	10763
Guards, hand, M.L.M. & M.E. rifles	Repair	11034
Guides, muzzle, Nos. 1 & 2	New patterns	14971
Haversacks	Obsolete	14575
Hook, suspending sword-bayonets	Introduction	10221
Implement, action, rifle, short, M.L.E. (Mk I)	Introduction	12057
Implement, action, pattern "D"	New pattern	10637
Implement, intrenching, P' 1908	Introduction	14796
Lance, exercise (Mk IV)	Alteration	10558
Lanyard, pistol, khaki (Mk IV)	New pattern	13058

Locket, union, Cadets (Mk II)	Introduction	11332
Lockets, union, Guards, Irish	Introduction	10472
Lockets, union, Royal Marines & Universal	Different design	11296
Lockets, union, Royal Marines & Universal	Introduction	10441
Loops, scabbard for P' 1888 scabbard	New patterns	11154
Mandril, scabbard, sword-bayonet, P' 07	Introduction	14676
Mandril, scabbard, sword-bayonet, P'07	New pattern	14755
Mandril, scabbard, sword, cavalry, P' 1885	Extended use	10123
Mandril, scabbard, sword, cavalry, P' 1908	Introduction	14712
Mineral jelly	Introduction	10131
Mineral jelly, zinc oil bottles	Continued use	10342
Mineral jelly, yellow	Obsolete	14209
Musket, fencing (Mk V)	Extended use	14756
Musket, fencing (Mk VI)	Introduction	11489
Musket, fencing (Mk VII)	Introduction	14757
Musket, fencing (Mk VII)	New pattern	14902
Musket, fencing, short (Mk I)	Introduction	12764
Musket, fencing, short (Mk II)	Alteration	14470
Musket, fencing, short (Mk II)	Introduction	13991
Muskets, fencing	Nomenclature	14903
Muskets, fencing	Obsolete	14903
Oil, petroleum, Russian, lubricating	Extended use	14209
Ornament, gilt, pouch, Conductors (Mk I)	Introduction	11080
Ornament, gilt, pouch, Engineers (Mk II)	Introduction	11112
Ornament, brass pouch, Dragoons	Obsolete	10561
Ornaments, Household Cavalry	Introduction	11203
Pistol, signal, Very's cartridge (Mk II)	Introduction	12741
Pistols, Webley, Mks I, I*, II & III	Repair	10190
Pistols, Webley, Mk III & IV	Requirement	13057
Plates, waistbelt, Artillery (Mk II)	Introduction	11113
Pouch, ammun., brown, pistol, "Sam Browne"	Introduction	10440
Pouch, ammun., brown, pistol, "Sam Browne"	Introduction	11267
Pouches, ammunition, black	New patterns	11037
Pouches, ammunition, buff	Obsolete	10561
Pouches, writing materials	New patterns	11037
Protector, front sight, M.M. & M.E.	Introduction	11714
Pull through, double (Mk I)	Withdrawn	12610
Rifle, charger loading, magazine, L.M. (Mk I)	Introduction	13992
Rifle, charger loading, magazine, L.E. (Mk I*)	Introduction	13992
Rifle, charger loading, M.L.M. (Mk II)	Alteration	14758
Rifle, magazine, Lee-Metford (Mks I & I*)	Sighting	10798
Rifle, magazine, Lee-Metford (Mk II)	Enfield barrel	12184
Rifle, magazine, Lee-Metford (Mks II & II*)	Backsight	10439
Rifle, magazine, Lee-Metford (Mks II & II*)	Sighting	10798
Rifle, short, magazine, Lee-Enfield	Alteration	13742
Rifle, short, magazine, Lee-Enfield (Mk I)	Alteration	13509
Rifle, short, magazine, Lee-Enfield (Mk I)	Introduction	11715
Rifle, short, magazine, Lee-Enfield (Mk I)	LoC cancelled	11947
Rifle, short, magazine, Lee-Enfield (Mk I*)	Introduction	13577
Rifle, short, magazine, Lee-Enfield (Mk I**)	Introduction	14936
Rifle, short, magazine, Lee-Enfield (Mk III)	Introduction	13853
Rifle, short, magazine, L.E., cond. (Mk I)	Introduction	11948
Rifle, short, magazine, L.E., cond. (Mk I)	New bolthead	12091
Rifle, short, magazine, L.E., cond. (Mk I)	Obsolete	13510
Rifle, short, magazine, L.E., cond. (Mk II)	Alteration	13509
Rifle, short, magazine, L.E., cond. (Mk II)	Introduction	11949

Rifle, short, magazine, L.E., cond. (Mk II*)	Introduction	13578
Rifle, short, magazine, L.E., cond. (Mk II**)	Introduction	14936
Rifle, short, magazine, L.E., cond. (Mk II***)	Introduction	14936
Rifle, short, magazine, L.E., cond. (Mk IV)	Alteration	14938
Rifle, short, magazine, L.E., cond. (Mk IV)	Introduction	13854
Rifle, short, M.L.E.	Alteration	14904
Rifle, short, M.L.E. (new & converted)	Spare bolthead	12611
Rifle, short, .22 RF	Introduction	14139
Rifles	Barrel marking	15000
Rifles	Nosecap	10687
Rifles	Shortened lead	14234
Rifles	Butt stock	11289
Rifles, charger loading	Extractor	14236
Rifles, charger loading	Butt stock	11289
Rifles, M.E., Mks I* & II*	Introduction	11810
Rifles, M.L.E., Mks I & I*	Altered sights	10371
Rifles, M.L.E., Mks I & I*	LoC cancelled	10393
Rifles, M.L.M., Mks II & II*	Enfield barrels	11498
Rifles, M.L.M. & M.L.E.	Piling swivel	13137
Rifles, short, M.L.E. A.T.	Introduction	13649
Rifles, short, M.L.E. A.T.	Nomenclature	14042
Rifles, short, M.L.M. & M.L.E. A.T.	Introduction	14168
Rifles, short, M.L.E.	Alteration	13313
Rifles, short, M.L.E.	Alteration	13549
Rifles, short, M.L.E.	Alteration	14937
Rifles, short, M.L.E.	Numbering	14070
Rifles, short, M.L.E., Mk III & cond. Mk IV	Extractor	14236
Rifles, short, M.L.E., Mk III & cond. Mk IV	LoC cancelled	14363
Rifles, short, magazine, Lee-Enfield (Mks I & cond. Mk II)	Butt trap	14362
Rifles, various	Alteration	14710
Rifles, .303-inch	Worn barrels	11033
Rod, cleaning, aiming tube	Introduction	13855
Sabretaches	Obsolete	10765
Saddlery, universal; frog, sword, saddle	Introduction	10444
Scbd, brown leather, sword-bayonet P' 1888	Nomenclature	10471
Scbd, brown leather, sword-bayonet P' 1888, land, Mk II	Introduction	11151
Scbd, brown leather, sword-bayonet P' 1888, naval, Mk I	Introduction	11109
Scbd, sword-bayonet, 1888	New pattern	10191
Scbd, sword-bayonet, P' 1888, Land, Mk II*	Introduction	12612
Scbd, sword-bayonet, P' 1903, Land, Mk I	Introduction	11811
Scbd, sword-bayonet, P' 1903, Naval, Mk I	Introduction	11812
Scbd, sword-bayonet, P' 1903, Naval, Mk II	Introduction	12185
Scbd, sword-bayonet, P' 1903, Naval, Mk III	Introduction	12568
Scbd, sword-bayonet, P' 1907 (Mk I)	Introduction	14169
Scbd, sword-bayonet, P' 1907 (Mk I)	Obsolete	14677
Scbd, sword-bayonet, P' 1907 (Mk II)	Introduction	14678
Scbd, sword, naval, P' 1900	New pattern	10419
Scbds, sword-bayonets	External chape	14866
Screw, keeper, striker, M.L.E. rifles, short	Introduction	13511
Screws, band & swivel	Alteration	14905
Slide, backsight, M.L.E. C.L. rifle, Mk I*	New pattern	14835
Slide, sight, wind-gauge	New pattern	11314
Sling, rifle, web, G.S. (Mk I)	Alteration	12060

Sling, rifle, web, G.S. (Mk I)	Introduction	10442
Swivel, band, Rifle, short, M.L.E.	Introduction	12992
Sword, Cavalry, Household, P' 1892 (Mk II)	Introduction	11290
Sword, Cavalry, P' 1908 (Mk I)	Introduction	14325
Sword, drummers (Mk II)	New pattern	11330
Sword, naval, P' 1900	Introduction	10419
Sword, practice, gymnasia, cond. 1895 Patt.	Introduction	10042
Sword, practice, gymnasia, P' 1904 (Mk I)	Introduction	12328
Sword, practice, gymnasia, P' 1907 (Mk I)	Introduction	13993
Sword, staff-serjeants, P' 1898	Introduction	11290
Sword, staff-serjeants, P' 1905 (Mk I)	Introduction	12825
Sword-bayonet, P' 1888 (Mk III)	Introduction	11151
Sword-bayonet, P' 1888 (Mks I, II & III)	Markings	11292
Sword-bayonet, P' 1903	Introduction	11716
Sword-bayonet, P' 1903, converted	Introduction	11717
Sword-bayonet, P' 1907 (Mk I)	Introduction	14169
Sword-bayonet, P' 1907 (Mk I)	Sharpening	15001
Swords (all patterns)	Buff pieces	10974
Swords (all patterns)	Sharpening	10559
Swords (all patterns)	Sharpening	12866
Swords, pioneers & buglers	Obsolete	12058
Swords, practice, mounted corps	Obsolete	14972
Tin, mineral jelly	Introduction	11950
Tin, mineral jelly	Use extended	12186
Tool, armourers— Gauges, striker point	Alteration	10638
Tool, clearing, .303-inch arms	Obsolete	10560
Tool, removing handguards, No. 1	Extended use	14043
Tool, removing striker, Rifle, short, M.L.E.	Introduction	13579
Tools, adjusting foresight, Rifle, C.L.	New pattern	14326
Tools, adjusting foresight, Rifle, short, M.L.E.	Introduction	12993
Tools, sdjusting foresight, Rifle, short, M.L.E.	New patterns	14326
Tools, armourers	New patterns	12422
Tools, armourers— Anvil, stock, butt	Introduction	11293
Tools, armourers—Anvil, stock butt, S.M.L.E.	Introduction	11951
Tools, armourers— Driver, screw, forked	For sword byt.	11294
Tools, armourers—Mandril, M.E. bayonet	Introduction	10394
Tools, armourers—Mandril, scabbards	Introduction	11295
Tools, armourers— Muzzle guides	Introduction	11079
Tools, armourers— Muzzle guides	New pattern	11851
Tools, armourers— Rod, cleaning	Introduction	11079
Tools, armourers— Rod, cleaning	New pattern	11851
Tools, armourers— Tool, sight protectors	Introduction	11718
Tools, armourers— Tool, sight line	Introduction	10688
Tools, armourers— Wrench, bolt head	Introduction	11035
Tools, sight line	New patterns	13994
Tube, aiming, .22-inch, M.L.M. rifle	Introduction	13856
Tube, aiming, .22-inch, M.L.M. & M.L.E. rifle	Introduction	13650
Tube, aiming, .23-inch, M.L.E. Rifle, short	Introduction	15002
Valise equipment, P' 1888	Modifications	13713
Valise equipment, P' 1888; Belt, waist	Introduction	10640
Valise equipment, P' 1888; Belt, waist	Modification	14945
Web equipment, P' 1908	Alteration	14687
Web equipment, P' 1908	New patterns	14288
Web equipment, P' 1908	Haversack cond.	14289
Web equipment, P' 1908, carriers, intrenching tool	Introduction	14795